Compression techniques for polymer sciences

Compression techniques for polymer sciences

Dr. Bradley S. Tice

WOODHEAD PUBLISHING INDIA PVT LTD

New Delhi

Published by Woodhead Publishing India Pvt. Ltd.
Woodhead Publishing India Pvt. Ltd.,
303, Vardaan House, 7/28, Ansari Road,
Daryaganj, New Delhi - 110002, India
www.woodheadpublishingindia.com

First published 2015, Woodhead Publishing India Pvt. Ltd.
© Woodhead Publishing India Pvt. Ltd., 2015

Woodhead Publishing India Pvt. Ltd. ISBN: 978-93-80308-31-9
Woodhead Publishing India Pvt. Ltd. e-ISBN: 978-93-80308-98-2

Typeset by Mind Box Solutions, New Delhi
Digitally Printed and bound by Replika Press Pvt. Ltd.

Contents

Dr. Tice is Institute Professor of Chemistry at Advanced Human Design located in The Central Valley of Northern California, U.S.A. Advanced Human Design is a research and development company with a focus on telecommunications. computing and material science. Dr. Tice is a Fellow of The Royal Statistical Society, a member of The Royal Pharmaceutical Society, and a Fellow of The British Computer Society.

Dr. Bradley S. Tice, FRSS & FBCS and a Member of The Royal Pharmaceutical Society

Photograph from the AIP Emilio Segre Visual Archives 2014

The monograph addresses the use of algorithmic complexity to perform compression on polymer strings to reduce the redundant quality while keeping the numerical quality intact. A description of the types of polymers and their uses are followed by a chapter on various types of compression systems that can be used to compress polymer chains into manageable units. The work is intended for graduate and post-graduate university students in the physical sciences and engineering.

Dr. Bradley S. Tice
FRSS & FBCS
September 2012

Introduction

The book examines algorithmic compression techniques for use on polymer chains. Because polymer chains are a linear sequence of units that have properties of redundancies of sub-groups of common or like-natured units that can be compressed to save space and present a perceptual change for analysis of the whole linear sequential segment. The author has previous publications on the area of algorithmic complexity and has addressed their use to polymers in this monograph (Tice, 2009 and 2010).

The book is arranged into chapters that address polymers, chemical processes, algorithmic complexity and then applied aspects of algorithmic complexity to polymer chains. Each chapter is a self-contained section that addresses that particular topic and lends itself well to understanding the application and theory of polymer compression techniques.

The book addresses large chain and multiple sequence polymers that are currently found in daunting amounts in 'Big Data' and a chapter and extensive appendix sections are added to address the 'real world' problems of massive data sets of polymer information.

Chapter 1
Polymers

1
Polymers

A polymer is a chemical material, or materials, that consists of repeating structural components that are formed through the process of polymerization (Wikipedia, 'Polymer', 2013: 1). The word 'polymer' comes from the Greek words 'poly' to mean 'many' and 'Zeros' to mean 'parts' (Borchardt, 1997: 1230).

Shapes of polymer chains, molecules linked together to form a sequence of connected molecules, have several types of geometries beyond the linear polymer chains (Hiemenz and Lodge, 2007: 7). Branched and cross-linked polymer chains are common and arise from the 'backbone' of a linear molecule (Hiemenz and Lodge, 2007: 7). The amount of such polymer branching structure is a branching upon branching of a molecule will result in a network type of geometry that is termed cross linked (Hiemenz and Lodge, 2007: 8). Some multi-branched molecules have discrete units and are termed hyper-branched polymers and other multi-branched polymers known as dendrimers, or tree-like molecules (Hiemenz and Lodge, 2007: 8).

Co-polymers are repeating units of polymers that have more than one type of repeating polymer unit and a polymer chain that has only a single type of repeating polymer unit is termed homo-polymers (Hiemenz and Lodge, 2007: 9). Also a co-polymer is a series of monomers that repeat in a chain and are bounded by each of their original monomer states (Hiemenz and Lodge, 2007: 9). The tertiary structure of polymer is the 'overall' shape of a molecule, and formal polymer nomenclature uses the structure of the monomer or repeat unit as a system of identification by the IUPAC or the International Union of Pure and Applied Chemistry (Hiemenz and Lodge, 2007: 18).

Monomers and repeat units of monomers are the primary descriptive quality of polymers and are categorized in the nomenclature according to the type of structures involved in the monomers (Wikipedia, 'Polymer', 2013: 6). Single type of repeating monomers are known as homopolymers, while a mixture of repeat monomers are known as co-polymers (Wikipedia, 'Polymer', 2013: 6).

The arrangement of monomers in co-polymers is as follows (Wikipedia, 'Polymer'. 2013: 9–10):

Alternating co-polymers:

–A–A–A–A–A–A–A–A–A–A–A–A–

Periodic co-polymers:

–A–B–A–B–A–B–A–B–A–B–A–B–

Statistical co-polymers:

–A–B–B–B–A–B–A–B–A–A–A–A–

Block co-polymers:

–B–B–B–B–B–B–A–A–A–A–A–A–

Graft co-polymers:

–A–A–A–A–A–A–A–A–A–A–A–A–

II

B–B–B–B–B–   –B–B–B–B–B

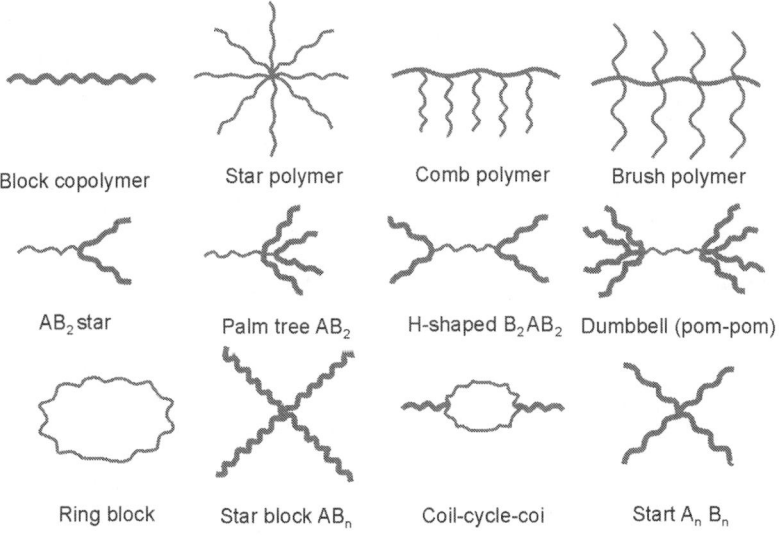

Various polymer architectures

Polymer architecture is the microstructure that develops the way the polymers branching points lead to a variation from a linear polymer chain (Wikipedia, 2013, "Polymer": 7). A branched polymer molecule is composed of a main chain of polymers, and one or more branched sub-chains that form a geometric list of branched polymer forms as follows (Wikipedia, 2013, "Polymer": 7–8).

The physical properties of a polymer are dependent on the size and length of a polymer chain (Wikipedia, 2013, "Polymer": 7–8) Polymerization is the length of the chain of polymers, and the physical size is also measured as a molecular weight.

Polymer	Important properties	Uses
Low-density polyethylene (LDPE)	Good toughness and pliability, transparent in films, excellent electrical insulator, resistant to chemicals	Films used in food wrapping and other applications, drapes, tablecloths, electrical wire and cable insulator, squeeze bottles

Contd...

Contd...

Polymer	Important properties	Uses
High-density polyethylene (HDPE	Higher crystallinity, soften¬ing temperature, hardness, and tensile strength than LDPE	Bottles (especially for liquid laundry detergent), coat-ings, pipes, wire and cable insulation
Polypropylene	Relatively low density; high tensile strength, stiffness, and hardness	Carpeting, injection-molded parts for appliances, photocopiers, and other machines
Polyvinyl chloride	Resistant to fire, moisture, and many chemicals, degraded by heat and ultraviolet light	Food containers, floor coverings, films, rainwear, coatings for electrical cables and wires
Polystyrene	Transparent and colorless, easily colored by pigments, good resistance to chemicals, good electrical insulator	Foam packing material, plastic parts, housewares, toys, packaging
Polytetrafluoroethylene (Teflon[1]" polymer)	Chemically inert, low friction properties	Coatings for cookware, insulation for wires, cables, motors, non-lubricated bearings

Contd...

Contd...

Polymer	Important properties	Uses
Polymethyl methacrylate	Transparent, colorless, high impact strength, poor abrasion resistance	Applications in which light transmission is needed: signal light lenses, signs, eyeglass lenses
Styrene-butadiene rubber	Equal or better physical properties compared with natural rubber except for heat resistance and resilience	Tire treads for cars (but not trucks, where natural rubber is used due to its better heat resistance and resilience)
Polyvinyl acetate	Water-sensitive, good adhesion	Water-based latex paints, to produce polyvinyl alco-hol, low-molecular-weight polymer in chewing gum
Polyvinyl alcohol	Water-soluble, fair adhesion	Water-thickening agent, packaging film
Polyvinyl butyral	Good adhesion to glass, tough, transparent, re-sistant to ultraviolet degradation	An inner layer in automo¬tive safety glass (wind-shields)

Products of polymers started in 1839 with Charles Goodyear with the mixing of sulfur and natural rubber to produce a useful product (Borchardt, 1997:1230). Some popular uses for polymers are industry-manufactured synthetic fibers produced by melted polymers and found in the following common fibers (Borchardt, 1997: 1234).

Polymer type	Comments	Typical uses
Nylon	amide group repeat units	carpeting, upholstery, clothing, tire cord
Acrylic	Acrylonitrile units constitute 90% or more of the polymer weight	carpeting, Drapes, blankets, sweaters, other articles of clothing
Polyester	Copolymer of a diacid and a diol (molecule with two hydroxyl groups)	Permanentpress clothes, underwear, 2-liter soda bottles
Spandex	Polyurethant	Athletic clothes, girdles, bras
Rayon	Made from cellulose	Clothing, blankets, curtains, tire cord

The examples used in this book are basic linear models of polymer chains to give emphasis on their application to the algorithmic complexity program.

The use of carbohydrates will be used as examples for compression purposes of algorithmic complexity especially various glucose models of carbohydrates. The following tables are 'glucose' tree models of two types of D-'glucose' polymers (Stick, 2001: 15–16).

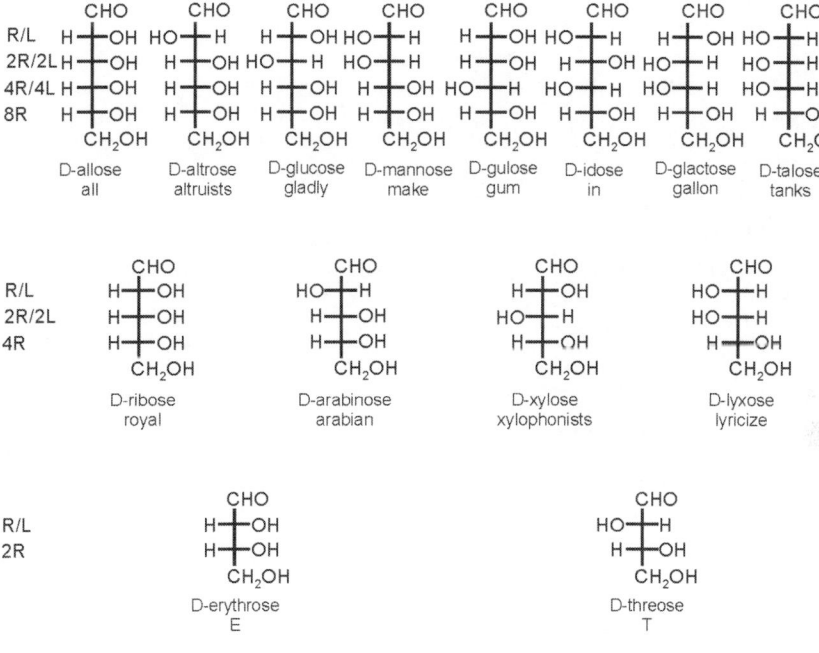

The D-family tree of the aldoses

	CH$_2$OH	CH$_2$OH	CH$_2$OH	CH$_2$OH
	CO	CO	CO	CO
R/L	H—OH	HO—H	H—OH	HO—H
2R/2L	H—OH	H—OH	HO—H	HO—H
4R	H—OH	H—OH	H—OH	H—OH
	CH$_2$OH	CH$_2$OH	CH$_2$OH	CH$_2$OH
	D-psicose pure	D-fructose fruits	D-sorbose sweetly	D-tagatose taste

	CH$_2$OH	CH$_2$OH
	CO	CO
R/L	H—OH	HO—H
2R	H—OH	H—OH
	CH$_2$OH	CH$_2$OH
	D-*erythro*-pent-2-ulose	D-*threo*-pent-2-ulose

	CH$_2$OH
	CO
R	H—OH
	CH$_2$OH
	D-*glycero*-terulose

CH$_2$OH
CO
CH$_2$OH

dihydroxyacetone

The D-family tree of the ketoses

Chapter 2
Compression of data

2
Compression of data

A mathematical notion of compression can be found in the fields of computer science and information theory (Wikipedia, 'Data Compression' 2013: 1–10). The use of fewer bits of information to represent the original code structure of information, or data, is termed either data compression, source coding or bit-rate reduction and is either a lossless or lossy compression format (Wikipedia, 'Data Compression', 2013: 1). Lossless compression is the reduction of statistically redundant information with no loss to the original information content (Wikipedia, 'Data Compression', 2013: 1). Lossy compression is the removal of unnecessary information from the original source, hence the term 'source' coding, with the resulting loss of the amount of original information or data (Wikipedia, 'Data Compression', 2013: 1).

Symmetrical data compression is when the time to compress is the same as decompression and asymmetrical data compression is when compression and decompression times vary (Wikipedia, 'Data Compression Symmetry', 2013: 1).

Universal code data compression is a prefix code that transposes positive integers within binary code words that are monotonic with relation to statistically probable distribution of lengths an optimal code would have produced (Wikipedia, 'Universal Code', 2013: 1). On average, prefix codes are assigned longer code words to larger integers, but are not used for precisely known statistical distribution and no universal code is known to be optimal for probability distribution (Wikipedia, 'Universal Code', 2013: 1). A universal code is not a universal source code as no fixed prefix code is used and both Huffman coding and arithmetic encoding are better at compression than universal coding, unless the exact probability of the message is not known and then a universal code is needed for compression (Wikipedia, 'Universal Code', 2013: 1–2).

Huffman codes were created in 1951 by David A. Huffman, a Ph.D. student at MIT, and are an entropy algorithm for encoding lossless data compression (Wikipedia 'Huffman Coding', 2013: 1). Huffman codes use a specific type of representing and choosing each symbol that is designated a prefix-free code so that any bit string represent a specific symbol is never the prefix of another symbol (Wikipedia, 'Huffman Coding', 2013: 1). There are many variations of Huffman codes (Wikipedia, 'Huffman Codes', 2013: 7–10).

Arithmetic coding, on average, have better compression values because of arbitrary number of symbols that can be combined for more efficient coding and function better when adapting to real world input statistics (Wikipedia, 'Huffman Coding', 2013: 1).

In entropy, encoding data compression is independent of the specific characteristics of the medium used and is a lossless compression code (Wikipedia, 'Entropy Encoding', 2013: 1). By assigning a unique prefix-free code to each specific symbol in the input and replace each fixed length input symbol with a variable length prefix-free output code word for compression (Wikipedia, 'Entropy Encoding', 2013: 1). The most common symbols use the shortest codes for compression (Wikipedia, 'Entropy Encoding', 2013: 1).

Decompression

Decompression is the state of returning a compressed sequence to its original size and sequential origins. There is exactness to this process of resurrecting a compressed sequential string in that both the accurate placement and type of symbol must occur to be the 100% realization of the original pre-compressed state of the sequential string.

A binary sequential string such as [11001100] is composed of two 1's in the initial position followed by two 0's and then two 1's with two 0's in the final two positions of this linear sequential string.

A compression of that binary sequential string, [11001100], would be as follows, [1010], reducing the total symbol quantity by half, from 8 symbols to 4 symbols, thus producing a successful compression of the original 8 symbol sequential string.

A de-compression of the compressed sequential string will produce the following key de-compression measures:

- That the initial, and following symbols, to the last, or terminal, symbol match the original sequential strings symbol placement and position within the totality of the original sequential string.

- That there are no more, and no less, than the exact number of symbols present in the pre-compressed original state sequential string.

An accurate and precise de-compression is mandatory for a useful application of this algorithmic complexity technique.

In simplest terms compression is a process of 'reducing' redundant data from an object. This redundant data is an original part of the object and so it must be said that it is an original feature of that object, even though it may seem excessive to the original design of that object. An example from genetics has redundant genetic instructions, hence the term 'junk' DNA, as being an important coding system for supporting healthy development of the organism (Princeton University, 2010: 1–3).

Take an object that is made up of a simple alphabet of four letters: ABCD; and that alphabet is used in a linear manner to function as a symbolic relationship to each other to form a strict hierarchy of spatial interaction in that the symbol A is always in the initial position always followed by the symbol B that is always followed by C with the symbol D always in the last position. The grammar of this ABCD alphabet would always be as follows: ABCD, with no variation permitted by the rules of concatenation by each symbol of that four letter alphabet. If this ABCD alphabet was allowed to reproduce itself, it would look like this arrangement: ABCDABCDABCDABCD, with the original alphabet being the initial four symbols: ABCD; and the duplication of three more sets of the original four symbol alphabet added in the suffix position.

A compressed version of this ABCD 'sentence', a group of words, symbols, unified to form a sentence, in this case the sentence being ABCDABCDABCDABCD, from the word ABCD. A compressed ABCD sentence would be [ABCD] plus ABCD added three times to the original ABCD word. The 4-symbol word was expanded into a 16-word sentence that could be compressed into a 4-symbol alphabet. De-compress the compressed ABCD sentence into a full 16-symbol ABCD sentence.

Compression should be seen as a 'reduction' of the original objects duplicate features, but not the reconfiguring of that original object total features. Compression and de-compression should be seen as mirror properties of the same object, nothing is lost, just hidden.

Chapter 3
Natural language compression

Natural language compression

Natural languages are those spoken by human beings and will be examined in this study as a 'written' language rather than the traditional verbal, or spoken, language.

Texting, the faddish uses of telephones, landlines, and mobile devices, cell phones, that use, in the English language system, 26 letters of the alphabet and 10 numerical symbols, an alphanumeric system, to send a message from one source to another source via a transmission system (Wikipedia, 'Texting Messaging', 2013: 1).

The act of compression is a natural act of parsimony of a sentence to reduce the amount of characters used without the loss of the 'content' of the message coded into this abbreviated form. Because the vowels of the English language – a, e, i, o, and u – are the connecting units of consonants, they form the 'significant' junction of words, especially prefix, suffix and, non-traditionally, infix positions in the English language, and by themselves, something English consonants do not normally do, become singularities of form and represent themselves accordingly.

Texting of the English language uses the 'dropping' of a vowel within an English word to save on space. The following example of such an English sentence is: 'Please pick me up at ten' can be shortened by dropping the 'ea' in Please, the 'ic' in pick and the letter 'ten' can be changed to the numerical 10 resulting in the following texting 'shorthand': 'Plse pk me up at 10' that has shortened the original 19 character message to the texting message's 14 character length.

Poetry is the most common form of language compression, both from a physical perspective, usually shorter than other forms of storytelling such as short stories, novels etc., and grammatically, made up of words and non-traditional or a 'slang' vernacular forms of speech that have a minimal word formation. Where poetry is also compressed is in the semantic value of the poem, in that fewer words are used to produce a more expanded notion of meaning, or meanings, than a traditional use of words and sentences.

Rewriting or editing a written work of literature is a form of compressing, in that how a word or sentence is composed can be duplicated in a shorter form by the addition of changing of words and sentences within a completed literary work. It has long been believed that short words are used more than long words due to George Zipf's 1930's work on frequency of words used by a speaker (MIT, 2011: 1–3). New research finds that the length of words is the amount of information the word contains, not the frequency of the use of that word (MIT, 2011: 1).

Even words that are misspelled in a sentence are 'automatically' corrected and interpreted as being a 'normal' spelling, but there is a limit on the number and types of 'jumbled' letters in a sentence for a native speaker to adjust to a readable pattern (ESRC, 2013: 1–3).

Number words have long had a place in man's history especially the number two, 2, that has the suggestive influence of being duel and mirror like in that human beings are bilateral and have two legs and two arms, two eyes, etc... and that common objects come in pairs (Menninger, 1969:12). The act of 'counting' as described by Otto Jesperson leads to objects as "more than one" without those objects being identical and that plurality presupposes difference as long as such differences have a conceptual common relationship at the core, such as a 'pair' and an 'apple', but not a 'brick and a castl e'(Jesperson,1924/1968:188).

The noted linguist Benjamin Lee Whorf (1897–1941) found that the Native American Indian language of the Hope Tribes use of verbs have a more precise and variant time and space correlation that European languages in that a physical object, when placed on a specific point, space, say a building on a fixed point of land, must be qualified as 'when'; time, the building was on that spot of land, and as such is a more perfect language for communicating space and time (Sebeok, 1966:580).

Whorf also studied the Hopi Noun system and found the descriptive qualities to have a large measure of variation. The concept of 'rain' is not just 'water falling from the sky', but is a descriptive quality of the 'properties' of the type of 'rain' being described such as: wet rain falling from sky, wet rain that runs in a river, frozen rain; snow, frozen rain pellets; hail, etc…, and that each of these physical actions are described by a separate noun formation (Carroll, 1956:140).

Numbers and words have a long history with both strong linguistic and mathematical lineage and the expansion and utilization of chemical and algorithmic properties adding to this long and developing history of man and his description of the world around him.

Chapter 4

Formal language compression

Formal language compression

Formal languages are computer languages and have their own rules and laws of formation, grammars, and connectivity, the structural qualities inherent in unions of parts of the whole. These are, by nature, linear and have the qualities of being mathematical in that they are represented as binary sequential strings.

Computers use formal languages that are limited to type of the input devices to input the information into the and the primary 'limiter' to this information is the computer key board, a modified version of the 'QWERTY' typewriter key board that was invented in 1878 with the popular success of the No. 2 Remington typewriter (Wikipedia, 'QWERTY', 2013: 1).

The language of mathematics is system of both natural languages, say the English language, that is invested with technical terms and grammar conventions to a specialized symbolic notation system (Wikipedia, 2013, "Language of mathematics": 1–8). The vocabulary of mathematics is by the use of notations that have grown over the years as mathematical developments have occurred. The grammar of mathematics is the specific use of a mathematical notation system, the rules that underlie 'how' and 'when' a particular notation is to be used with the framework of a mathematical discipline (Wikipedia, 2013, "Language of mathematics": 4). An example would be the English word formula one plus one equals two that can be mathematically notated as $1 + 1 = 2$. A hybrid of the two language systems, natural language and mathematical formalisms such as the word formula for one plus one equals two can be written as $1 +$ one equals 2.

The author has done research in the area of natural language and mathematics with a work addressing Kurt Godel's Continuum Theory use of word problems as proof of inconsistency in David Hilbert's axiomatic laws of mathematics. The author changed some of the words found in Godel's original paper that, in turn, changed the semantic nature of the sentences used to proof the axiomatic truths found in Godel's paper and by so doing changed the valuations used to determine an inconsistency (Tice, 2013).

Current research has found that a similarity in genetic codes and computer codes arises when the most frequently used of components of either a biological or computer system have the most 'descendants' (Brookhaven National Laboratory, 2013: 1). Crucial parts of a genetic code in the metabolic process of 500 bacteria species was measured against the frequency of 200,000 packages of an open source software program, Linux, and that the more a code, either biological or computer, it is adopted more often and takes up a greater portion of the system (Brookhaven National Laboratory, 2013: 2).

The rules of a formal language have specific grammar that gives clues to a message, or signal, being evaluated from outside the intended communications channel. In long series of passwords, coded words or properties of multiple words used to allow access to a secure message, that are based on words or phrases; sentences have a distinct 'grammar' pattern that can be 'de-cyphered' more accurately than a more random sequence of symbols. The longer the sentence, the easier it is to find the frequency of common usage in that some aspects of grammar, such verbs and nouns are more prevalent than other aspects of a language's grammar pattern (Carnegie Mellon University, 2013: 1).

Niels Jerne, in his Nobel Prize Lecture, compares the human body's immune system to 'inherent' traits of human language acquisition and that the 'seeds' of this genetic transference is in the 'DNA' of the human progenitors of the species (Jerne, 1984: 211–225).

While this book addresses carbohydrate molecules as sources of polymer chains, the 'chains of life', or DNA, are genetically encoded polymers containing life's language for continuing life. These genetic codes have set rules for functioning and become either viable or non-viable mutations when changes occur in the 'processes' of encoding or decoding a specific point within a genetic code.

While this book does not utilize genetic codes for either theoretical or applied aspects of algorithmic complexity programs, their application to both theoretical and applied genetics is for future publications and while some papers on theoretical genetics are included in the appendix section of this book, a major work on the subject must wait for future research papers and monographs.

Chapter 5

Types of compression program

Types of compression program

Entropy coding is used in computer science and information theory in the form of Huffman coding that produces lossless data compression (Wikipedia "Huffman coding", 2013: 1). Huffaman codes and 'prefix' codes are synonymous as Huffman codes are ubiquitous in computing.

The type of compression techniques to be examined in this book are algorithmic compression techniques as developed by the author and address both whole segments of common, or like-natured, polymer units and specific, or truncated, aspects of a linear sequential string or a random or non-random type.

A non-random binary sequential string is as follows:

[11001100110011]

A series of alternating binary characters sub-grouped into 2 like-natured characters that have a total length of 14 binary characters. A random binary sequential string is as follows:

[100011011000001111]

This is a pattern-less series of alternating 18 binary characters in length that are not compressible by traditional measures of compression. A compressible random binary sequential string is introduced in the next chapter and was discovered by the author in 1998.

Chapter 6

Algorithmic compression

Algorithmic compression

Algorithmic complexity is a sub-field of Claude Shannon's Information Theory (1948) and is known by various names as Algorithmic Information Theory, Kolmogorov Complexity, and Algorithmic Complexity to name a few such monikers and was developed by R.J. Solomonoff, A.N. Kolmogorov, and G.J. Chaitin in the 1960s (Chaitin, 1982: 38).

The traditional definition of Algorithmic Compression is the degree of difficulty, or complexity, needed for an object, usually a binary sequential string, to calculate or 'construct' itself (Chaitin, 1982: 38). On a perceptual level, this notion of complexity is associated with degrees of measurable 'randomness' found in a binary sequential string and is considered a form of Martin-Lof randomness because of the visible nature of the measure of randomness (Chaitin, 1982: 40).

This 'visible', or perceptual, nature of non-randomness is found in the regularity of a binary system, either 1's or 0's, in that a regular pattern of 1's and 0's would look like: 1010101010 or five 1's followed each by five 0's in a regular pattern of alternating symbols for a total of ten characters. A 'random', or irregular pattern, would look as follows: 1001110110 or a ten character total with a concatenated sub-grouping of [1] [00] [111] [0] [11] [0]. The significance of this finding is that a measurable nature of randomness could be described in statistics for the first time and set a defined boundary between randomness and non-randomness. This measure of randomness and non-randomness also had the properties of compressibility and non-compressibility of a binary sequential string.

The author discovered in 1998 a 'compressible' random binary sequential string and stands as the most precise and accurate measure of randomness known in statistical physics (Tice, 2009 and 2012). A compressible random binary sequential string program utilizes a traditional binary random sequential string such as [001110110001110] that is 15 characters in length and sub-groups each like-natured sub-group of either 0's or 1's as follows:

[00]+[111]+[0]+[11]+[000]+[111]+[0]

A notation system is used behind each initial character to give a total number of like-natured characters in each sub-group while removing the remaining like-natured characters from each sub-group as follows:

0 (2) +1 (3) +0+1 (2) 0 (3) +1 (3) +0

The compressed state of the original random 15 character binary sequential string is as follows:

0101010

The compressed state is 7 characters in length from the original pre-compressed state of a 15 character random string. Traditional literature on random binary sequential strings has this random 15 character binary sequential string as being unable to compress. The author's method of compression reduces, compresses, the original 15 characters into less than half of the original total length.

The author has expanded the use of radix-based number systems beyond the traditional binary, or radix 2 based number system, including two-digit radix numbers – 10, 12 and 16 – as well as radix 3, 4, 5 and 8 based number systems (Tice, 2013). These larger radix-based number systems can be more easily used with ternary and larger symbol systems that use more than a dual, or binary, system of characters. Analog systems are usually associated with more than two types of features, such as the four bases of the genetic code, that make it an ideal system for a radix 4 based number system.

While the symbols used for the characters are Arabic numbers, the use of [1] and [0] are to note contrasts of the [1] and the [0] Arabic symbols and not the quantitative numeration of a one [1] or a zero [0] number. The use of other symbols to represent the contrastive nature of the [1] and the [0] could be an alphabet that could use the letter [A] for [0] and the letter [B] for [1]. This minimalism is to reduce the symbol to a two-natured contrast of black and white, night and day, up or down type of referencing that denoted the opposite nature of each symbol in use.

Linguistics uses many binary systems in giving a 'scientific' description of a language process. Phonology, the study of sound patterns of spoken human languages uses a contrastive feature of a phonological property of a specific sound per single or two adjoined letters of an alphabet to assign a specific sound. When these letters make up a word, the letters in that word make up either a consonant or a vowel. The following phrase, a phrase is a spoken sentence, "The chemist is a scientist" can be written as a consonant and vowel representation of the alphabetical letters as sounds of that phrase. The phrase "The chemist is a scientist" can be written as "ccv ccvcvcc vc v ccwccvcc" with the c = consonant and v= vowel. This is a binary representation of a consonant and vowel model of the phonological pattern of an English language phrase as written as an English language sentence.

Because the areas of study are of a finite length, the techniques employed with involve either the whole finite string of polymers or subsections of that whole finite string of polymer chains. The use of an algorithmic complexity program to 'compress' only a section of a finite length of a polymer string can be applied in the same manner as the whole string. Each subsection of a whole finite string is made of common components of that whole finite string.

An example would be the following random binary sequential string of a twenty character length:

[11000100000111000110]

If the desired compression area within this whole finite string was the unit of [0] that comprised five sequential [0] symbols, then the seventh symbol, from the left initial side position, [0] and the sequential 4 [0's] following that seventh symbol are compressed a [0]x 5 as a notation that five sequential [0]symbols have been compressed and they can be represented as follows:

[110001{0}111000110]

Notice that brackets have been used to note the placement and symbol type of the 'compressed' 5 [0] section of the finite string.

The remaining 16 character length finite sequential binary string is reduced from the whole finite sting of the original 20 character string and reduced by four characters.

De-compression of the compressed section to its original placement and character type results into an exact reproduction of the original whole finite string.

Algorithmic complexity is a sub-field of Claude Shannon's Information Theory that was published in 1948 and has been adopted for use in chemistry in two monographs (Eckschlager and Danzer, 1994; Eckschlager and Stepanek, 1979). Both Information Theory and Algorithmic Information Theory have been evaluated on 'esthetic' principles as published by Moles (1966) and Stiny and Gips (1978).

Chapter 7
Chemical formulae

Chemical formulae are a way to express the number of atoms that make up a particular chemical compound using a notation system made up of a single line of chemical element symbols, numbers, and other symbols and are limited to a single typographic line (Wikipedia, 'Chemical Formula', 2013: 1).

Molecular formulae represent the number of each type of atom in a molecule of a molecular structure (Wikipedia, 'Chemical Formula', 2013: 1). Empirical formulae are the letters and numbers indicating atomic proportional ratios of one type of atom to another type of atom (Wikipedia, 'Chemical Formulas', 2013: 1).

Chemical formulae can be written as follows (Cartage.org., 'Structural Formulas', 2013: 1):

Common name:	**Formaldehyde**
Molecular formula:	CH_2O
Lewis formula:	H
	H:Ö:H
	Ḣ
Kekule formula:	H
	H
	I
	H–O–H
	I
	H

The importance of chemical nomenclature is that the 'language' of chemistry is by way of signs and symbols that denote a specific property or action within a chemical structure. This monograph uses Fischer Projections of carbohydrates and linear notation systems for chemical formulae as examples for compression of polymer molecules.

Herman Emil Fischer developed the Fischer projection in 1891 as a two-dimensional representation of a three-dimensional organic molecule usually a carbohydrate (Wikipedia, 2014, 'Fischer Projection': 3). A Fischer projection will give the impression that all horizontal bonds project toward the viewer and the vertical bonds project away from the viewer (Wikipedia, 2014, "Fischer Projection": 1).

Fischer projections are used to depict carbohydrate molecules and give a differentiation between L- and D- molecules with D-sugar carbon units having hydrogen on the left side and hydroxyl on the right side of the carbon backbone (Wikipedia, 2014, "Fischer Projection": 1). L-sugar carbon units will have hydrogen on the right side and hydroxyl on the left (Wikipedia, 2014, "Fischer Projection": 1). Fischer projections are used in biochemistry and organic chemistry to represent monosaccharide and amino acids as well as other organic molecules (Wikipedia, 2014,"Fischer Projection": 2).

The use of existing chemical nomenclature is to facilitate the ease of applicability, both theoretical and applied, to the use of an algorithmic complexity program for the polymer sciences.

Chapter 8
Fischer projection

Fischer projection

A Fischer projection is a two-dimensional depiction of a three-dimensional organic molecule by way of visual projection (Wikipedia, 'Fischer Projection', 2013: 1). Hermann Emil Fischer developed this graphing system in 1891 (Wikipedia, Fischer Projection', 2013: 1). The characteristics of Fischer projections are one or more stereogenic centers surrounded by four plain bonds aligned in a vertical fashion (Brecher, 2006: 1933). These Fischer projections were originally designed to depict carbohydrates (Brecher, 2006: 1933).

In Fischer projections, all bonds are represented as horizontal or vertical lines and are the only representational configurations that are not invariant with respect to rotation of depiction (Wikipedia, 'Fischer Projection', 2013: 1 and Brecher, 2006: 1934).

Both L-sugars and D-sugars are the 'mirror' image of the other with the D-sugars occurring naturally (Rensselaer Polytechnical Institute, 011: 2). Haworth projections are similar to Fischer projections and are used to depict sugars in ring form and Newman projections are used to represent stereochemistry of alkanes (Wikipedia, 2014,"Fischer Projection": 3).

Examples of Fischer projections (Rensselaer Polytechnic Institute, 2011: 1—2).

Monosaccharide:

Aldoses (e.g., glucose) have a aldehyde at one end.

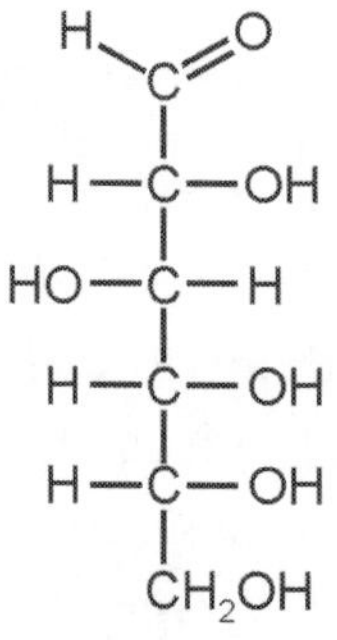

Ketoses (e.g., fructose) have a keto group, usually at C #2.

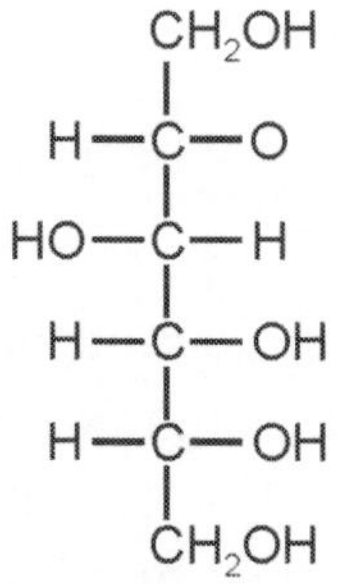

$$
\begin{array}{c}
H\!\!\diagdown_{\displaystyle C}\!\!\diagup^{O} \\
H\!-\!C\!-\!OH \\
HO\!-\!C\!-\!H \\
H\!-\!C\!-\!OH \\
H\!-\!C\!-\!OH \\
CH_2OH
\end{array}
$$

D-glucose

$$
\begin{array}{c}
CH_2OH \\
H\!-\!C\!-\!O \\
HO\!-\!C\!-\!H \\
H\!-\!C\!-\!OH \\
H\!-\!C\!-\!OH \\
CH_2OH
\end{array}
$$

D-fructose

$$
\begin{array}{c}
O\!\!=\!\!C\!\!-\!H \\
H\!-\!C\!-\!OH \\
HO\!-\!C\!-\!H \\
H\!-\!C\!-\!OH \\
H\!-\!C\!-\!OH \\
CH_2OH
\end{array}
$$

D-glucose

$$
\begin{array}{c}
O\!\!=\!\!C\!\!-\!H \\
HO\!-\!C\!-\!H \\
H\!-\!C\!-\!OH \\
HO\!-\!C\!-\!H \\
HO\!-\!C\!-\!H \\
CH_2OH
\end{array}
$$

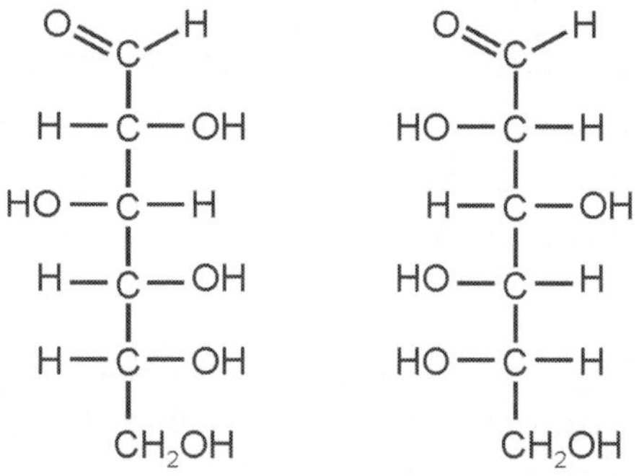

L-glucose

Chapter 9

Compression of polymers

The notion of algorithmic complexity as it relates to a series of symbols in a linear pathway as found in the chemical depictions of Fischer projections is the 'compression' of linear sequential similar symbols for reduction to a single symbol with a corresponding notation to denote the total number of characters of a similar type that were in the original chemical formula.

Fischer projections of carbohydrates

Monosaccharides are simple sugars with multiple hydroxyl groups (Molecular Biochemistry 1, 2013: 1).

Monosaccharides (Wikipedia, '2-Carb-3 Carb and 2-Carb-4', 2013: 3–4).

D-Glucose   CHO–HCOH–HCOH–HCOH–HCOH–CH$_2$OH

A compressed version of D-glucose would have the 4 HCOH units notated as HCOH(4) and would be written as follows:

D-Glucose (Compressed) CHO–HCOH(4)–CH$_2$OH

D-Xylose   CHO–HCOH–HCOH–HCOH–CH$_2$OH

A compressed version of D-Xylose would have the 3 HCOH units notated as HCOH(3) and would be written as follows:

D-Xylose (Compressed)   CHO–HCOH(3)–CH$_2$OH

D-arabino-Hex-2-ulose   CH$_2$OH–C=0–HOCH–HCOH–

(D-Fructose)   HCOH-CH$_2$OH

A compressed version of D-arabino-Hex-2-ulose would have the 2 HCOH units notated as HCOH(2) and would be written as follows:

D-arabino-Hex-2-ulose $CH_2OH–C=0–HOCH–$

(Compressed) $HCOH(2)–CH_2OH$

D-glycero-gulo-Heptose $CHO–HCOH–HCOH–HCOH–$

                             $HCOH–HCOH–CH_2OH$

A compressed version of D-glycero-gulo-Heptose would have the 5 HCOH units notated as HCOH(5) and would be written as follows:

D-glycero-gulo-Heptose $CHO–HCOH(5)–CH_2OH$

(Compressed)

L-Arabinose $CHO–HCOH–HOCH–$

                             $HOCH–CH_2OH$

A compressed version of L-Arabinose would have the 3 HCOH units notated as HCOH(3) and written as follows:

L-Arabinose $CHO–HCOH(3)–CH_2OH$

(Compressed)

L-glycero-D-manno-Heptose $CHO–HOCH–HOCH–HCOH–$

                             $HCOH–HOCH–CH_2OH$

A compressed version of L-glycero-D-manno-Heptose would have the 2 HOCH and 2 HCOH units notated as HOCH(2) and HCOH(2) and written as follows:

L-glycero-D-manno-Heptose $CHO–HOCH(2)–HCOH(2)–$

(compressed) $HOCH–CH_2OH$

Chapter 10

Line notation systems and compression

Line notation systems and compression

Line notation systems are used as a linear notation system to specify sub-structural patterns and structural patterns on chemical molecules (Wikipedia, 'Smiles arbitrary target specification', 2013: 1 and Wikipedia, 'Simplified molecular-input line-entry system', 2013: 1). Two systems are used by various chemical organizations: SMILES – Simplified Molecular-Input Line-Entry System and SMARTS – Smiles Arbitrary Target Specification (Wikipedia, 'Smiles...', 2013: 1 and Wikipedia, 'Simplified..', 2013: 1).

Compression of Line Notation Systems is done at the point of repetition of a chemical symbol that makes it 'redundant' in the formula's notation.

In the following notation for SMARTS, a recursive SMART notation is used to combine acid oxygen and tetrazole nitrogen to define oxygen atoms that are anionic (Wikipedia, 'SMARTS...', 2013: 4).

SMARTS Notation: [$([OH][C,S,P]=0),$([nH]lnnncl)]

The obvious compression point is the three sequential 'n' symbols [nnn] that can be compressed by either a numerical value for the 'total' number of n's such as n3 or by notating with an underlining of a single 'n' character to define a 'quantity' of three 'n' symbols: [nnn] = [n]. Chromatic properties of notations, such as monochrome, black, or chromatic, colored, notation symbols can be used to signal 'compression' points such as the previous example,

[nnn], with the 'n' symbol have a color other than the traditional black hue to signify three n's [nnn].

Using the SMILES notation system, the following SMILES notation for cyclohexane is as follows (Wikipedia, 'SMILES...', 2013: 4):

SMILES Notation: C1CCCCC1

The compression point is the five sequential C's [CCCCC] following the 1 in the formula. Again, a notational quantity of 5 C's can follow an initial C character to denote the number of C's symbols in the notation: C5 = [CCCCC]. The use of an underlined C character could also be used to define the number of C's in the formula: C = [CCCCC].

Chapter 11

Current trends in research

Current trends in research

In reviewing the current literature on compression of polymers, the following two types of areas appear: general polymers and bioprocess polymers, genetics. General polymers are all non-genetic related polymers and bioprocess polymers are genetic oriented polymers for biological systems.

A sampling of the current literature on compression of genetic codes, DNA and RNA, has resulted in new developments in the analysis and use of genetic codes to science and engineering. Bioprocess polymers have advanced to the point of monographs being written in this area of science with the topics of DNA computing and algorithmic developmental processes in nature (Condon et al., 2009).

General polymers

Alagoz (2010) studied the effects of sequence partitioning on

compression rates and found to have a greater compression rate on a sequence (Alagoz, 2010: 1). The University of Wisconsin-Madison (2010) notes that self-assembling polymer arrays improve data storage potential with the use of block co-polymers (University of Wisconsin-Madison, 2010: 1). In *The Journal of the American Chemical Society* (2010), a paper notes that an organic ternary data storage device was developed using a three value system (J. ACS, 2010, 1). Evans et al. (2001) used a symbol compression ratio for string compression and estimation of Komogorov complexity (Evans et al., 2001: 1).

Biopolymers

Svoboda (2010) DNA sequencing is done with a DNA 'transistor' (Svoboda, 2010: 1). In the Physical Review of Letters (1998), a paper notes the use of the entropic segmentation method to define the sequence compositional complexity of DNA (Phy. Rev. Lett., 1998: 1). Palmer (2011) uses a DNA computer to calculate square roots (Palmer, 2010: 1). Pollack (2011) notes the huge amount of DNA sequencing is producing a large amount of data (Pollack, 2011: 1). Cherniavsky and Ladner (2004) study grammar-based compression of DNA sequences (Cherniavsky and Ladner, 2004: 1).

TGen (2010) uses an encoding sequencing method to compress genomic sequencing data (TGen, 2010: 1). The University of Reading (2010) have created a synthetic form of DNA that functions as a linear sequence of letters for use in information technology (Phys. org, 2010: 1). Inderscience (2009) notes the use of a computer data base to compress DNA sequences used in medical research (Science Daily, 2010: 1). Condon et al. (2010) has published a book on algorithmic bioprocesses that addresses organic compression of biopolymers (Condon et al., 2010).

Chapter 12
Big data

Big data

Big data is the term that describes the ever growing amount of 'data' that is being collected by various disciplines, most notably genetics. This 'glut' of data has little form to make it a 'content' and 'contextual' value of 'information' and is more a gigantic pattern of 'excess' rather than a approachable part of an emerging whole.

The use of algorithmic complexity programs for compression of data streams and theoretical uses in modeling chemical and biological processes is a major factor in managing Big Data and giving a functional tool for addressing the major problem with large data pools: too much 'data' and too little 'information'.

Also the type of 'data' derived from new techniques has to be weighed against existing 'information' on the resulting study. A current research project reported that a series of internal points, or nodes, in an organism have a great hierarchical value to the 'operation' of that organism (Northwestern University, 2013: 1). These 'necessary' nodes can descript the whole process without 'monitoring' the 'actions' of the components of that organism (Northwestern University, 2013: 2).

Similar findings about 'internal' connectivity in organisms and the 'behavior' of such organisms was clearly detailed the various 'feedback' theory groups of the 1940s and 1950s, including Wieners 'Cybernetics', and to a lesser degree, Shannon's Information Theory, not to mention Canon's concept of 'homeostasis' in living systems that was done in the 1920s (Hayes, 2011: 78–81). Much of these 'new' research programs are just extensions of older studies, such as 'systems theory' such as Ludwig von Bertalanffy's 'General System Theory' (1928) for biological processes (Wikipedia, 'Systems Theory', 2013: 3–4). Systems theory seems to be a subset of cybernetics theory in that cybernetics was engineering related and was a feedback-oriented system that was based on external interacting feedback loops that produced finite results (Wikipedia, 'Systems Theory', 2013: 8). Bertalanffy's work is pre-dated by the Russian Alexander Bogdanov's three volume work that was published in Russia between 1912 and 1917 (Wikipedia, 'Systems Theory', 2013: 8).

My research into feedback systems can be found in my unpublished manuscript titled *Modeling Complexity in Molecular Systems: A Revised Edition* (2013) that uses various techniques from the engineering field to feedback systems found in biological processes (Tice, 2013). These techniques, along with some advances in development in graph theory by the author, have added significant methodology to the analysis and modeling of aspects to Big Data.

By using such effective techniques as tools for managing and the analysis of Big Data can the situation be addressed in a productive manner without falling into the void of 'unproductive' data collections (Ovide, 2013: 1).

Conclusion

Conclusion

The book has presented tools and techniques in the form of algorithmic compression programs to compress all or part of common, or like-natured, units found in polymer chains. Both random and non-random sequential strings have been examined in polymer sequences regarding the processes of compression.

Summary

Summary

In reviewing the material for this book, I gave emphases on the basics of polymers and the use of algorithmic complexity as a compression technique. The use of such tools to large data pools of DNA sequences is obvious and I expect to see many advancements over the next 30 years in this area of information management of data.

References

References

Alagoz, B.B. (2010). "Effects of sequence partitioning on compression rates". November 3, 2010, pp. 1–6. Website: http://arxiv.org/abs/1011.0338

Brecher, J. (2006). "Graphical Representation of Stereo-chemical Configuration". Pure and Applied Chemistry, Volume 78, No. 10., pp. 1897–1970.

Brookhaven National Laboratory (2013). "Researchers find surprising similarities between genetic and computer codes". Phys. org. March 29, 2013, pp. 1–3.
http://phys.org/news/2 013–0 3-similarities-genetic-codes-html

Carnegie Mellon University (2013). "Grammar undercuts security of long computer passwords". ScienceDaily. January 27, 2013, pp. 1–3. Website:
http://www.sciencedaily.com/releases/2013/01/13012412354 9.html

Carroll, J. B. (1956). Language, Thought, and Reality. Cambridge: The MIT Press.

Cartage.org. (2013). "The importance of structural formulas".March 16, 2013, pp. 1–3. Website:
http://www.cartage.oeg.lb/en/themes/Sciences/Che,istry/Organi...

Chem.qmul. (2013). "2-Carb-3 and 2-Carb-4". March 16, 2013, pp. 1–5. Website:
http://www.chem.qmul.ac.uk/iupac/2carb/033n04.html.

Cherniavsky, N. and Ladner, R. (2004). "Grammar-based compression of DNA sequences". UW CSE Technical Report 2007–05–02. May 28, 2004, pp. 1–21.

Condon, A, Harel, D., Kok, J.N., Salomaa, A. and Winfree, E. (2010). Algorithmic Bioprocesses. Berlin: Springer Publishing.

Eckschlager, K. and Stepanek, V. (1979). Information Theory as Applied to Chemical Analysis. New York: John Wiley & Sons.

*Eckschlager, K. and Danzer, K. (1994). Information Theory in Analytical Chemistry. New York: John Wiley & Sons.

EMBL (2013). "Researchers make DNA data storage a reality: every film and tv program ever created-in a teacup". ScienceDaily, January 24, 2013, pp. 1–3. Website:
http://www.sciencedaily.com/2013/01/130123133432.htm

ESRC (2013). "How can we still raed words when the lettres are jmbbuled up?". ScienceDaily, March 17, 2013. Website:
http://www.sciencedaily.com/releases/2013/03/130315074613.htm

Evans, S.C. and Bush, S.F. (2001). Symbol compression ratio for string compression and estimation of Kolmogorov complexity". CiteSeerX. September 3, 2010, p. 1. Website:
http://citeseerx.ist.psu.edu/viewdoc/summary?doi=10.1.21.8207

Hiemenz, P.C. and Lodge, T.P. (2007). Polymer Chemistry. 2nd ed. Boca Raton: CRC Press.

Inderscience (2009). "Computer database compresses DNA sequences used in medical research". ScienceDaily. July 9, 2010, pp. 1–2. Website:
http://www.sciencedaily.com/releases/2009/11/09111120105.html

Jern, N.K. (1984). "The generative grammar of the immune system". Nobel Lecture, December 8, 1984, 211–225.

Jesperson, O, (1924/1968). The Philosophy of Grammar. London: George Allen & Unwin, LTD.

J. ACS (2010). "Organic ternary data storage device developed". Journal of the American Chemistry Society. September 3, 2010, pp. 1–2. Website:
http://www.physorg.com/print190451302.html

Lagowski, J.J. (1997). MacMillan Encyclopedia of Chemistry. Volume 3 New York: MacMillan Simon & Schuster.

Menninger, K. (1969). Number Words and Number Symbols: A Cultivated History of Numbers. Cambridge: The MIT Press.

MIT (2013). "What determines the length of words? MIT researchers say they know". Phys. org. February 18, 2011, pp. 1–3. Website:
http://www.physorg.com/print21654 54 7 5.html

Moles, A. (1966). Information Theory and Esthetic Perception. Urbana: University of Illinois Press.

Palmer, J. (2011). "DNA computer calculates square roots". BBC News. June 2, 2011, pp. 1–3. Website:
http://www.bbcco.uk/news/science-environment-13626583?print=true

Physical Review of Letters (2010). "Sequence compositional complexity of DNA through an entropic segmentation method". October 15, 2010, pp. 1–2. Website:
http://prl.aps.org/abstract/PRI/v8 0/i6/pl34 4 1

Pollack, A. (2011). "DNA sequencing caught in deluge of data". The New York Times. December 1, 2011, pp. 1–5. Website:
http://www.nytimes.com/2011/12/01/business/dna-sequencing-caught=in-deluge-of-data.html

Princeton University (2010). "Redundant genetic instructions in "junk" DNA support healthy development". Science News, July 16, 2010, pp. 1–3. Website:
http://esciencenews.com/articles/2010/07/16redundant-genetic-instruetions-junk-dna-suppor

RPI.edu (2013). "Carbohydrates-Sugars and Polysaccharides". March 16, 2013, pp. 1–14. Website: http://rpi.edu/dept/bcbp/molbiochem/MBWeb/mbl/part2/sugar.hlmt

Sebeok, T.A. (1966). Portraits of Linguists: A Biographical Source Book for the History of western Linguistics 1746–1963. Bloomington: Indianan University press.

Stanford University Medical Center (2013). Biological transistor enables computing within living cells". ScienceDaily, March 9, 2013, pp. 1–5. Website: http://www.sciencedaily.com/releases/2013/03/1303281424 00.htm

Stick, R. V. (2001). Carbohydrates: The Sweet Molecules of Life. New York: Academic Press.

Stiny, G. and Gips, J. (1978). Algorithmic Aesthetics; Computer Models for criticism and Design in the Arts. Berkeley: University of California Press.

Svoboda, E. (2010). "The DNA transistor". Scientific American. December 2010, p. 46.

TGen (2010). "New technology reduces storage needs and costs for genomic data". ScienceDaily. July 9, 2010, pp. 1–3. Website: http://www.sciencedaily.com/releases/2010/07/100706150614.html

Tice, B.S. (2009). Aspects of Kolmogorov Complexity: The Physics of Information. Denmark: River Publishers.

Tice, B.S. (2012). A Level of Martin-L of Randomness. New Hampshire: Science Press.

Tice, B.S. (2013). Language and Godel's Theorem: A Revised Edition. Denmark: River Publishers.

University of Hertfordshire (2013). "New method to measure the redundancy of information". ScienceDaily. March 19, 2013, pp. 1–3. Website: http://sciencedaily.com/releases/2013/02/130214132809.htm

University of Reading (2010). "Genetic inspiration could show the way to revolutionise information technology". May 28, 2010, pp. 1–2. Website:
http://www.physorg.com/print196961035.html

University of Wisconsin-Madison (2010). "Self-assembling polymer arrays improve data storage potential". Esciencenews. September 3, 2010, pp. 1–2. Website:
http://esciencenews.com/articles/2008/08/14/self-assembling. polymer.arrays.improves.data.st...

Wikipedia (2013). "Big Data". March 18, 2013, pp. 1–8. Website: http://en.wikipedia.org/wiki/Big_data

Wikipedia (2013). "Systems theory". March 18, 2013, pp. 1–13. Website:
http://en.wikipedia/wiki/General systems theory

Wikipedia (2013). "Text messaging". March 16, 2013, pp. 1–21. Website:
http://en.wikipedia.org/wiki/Texting

Wikipedia (2013). "Smiles arbitrary target specification". March 16, 2013, pp. 1–7. Website:
http://en.wikipedia.org/wiki/Smiles arbitrary target specification

Wikipedia (2013). "Simplified molecular-input line-entry system". March 16, 2013, pp. 1–12. Website:
http://en.wikipedia.org/wiki/Simplified molecular-input line-ent

Wikipedia (2013). "QWERTY". March 16, 2013, pp. 1–15. Website: http://en.wikipedia.org/wiki/QWERTY

Wikipedia (2013). "Fischer Projections". March 16, 2013, pp. 1–3. Website:
http://en.wikipedia.org/wki/Fischer projection

Wikipedia (2013). "Chemical formula". March 16, 2013, pp. 1–11. Website:
http://en.wikipedia.org/wiki/Chemical formula

Wikipedia (2013). "Algorithmic Information Theory". May 2, 2013, pp. 1–7. Website:
http://en.wikipedia.org/wiki/Algorithmic information

Wikipedia (2013). "Huffman Coding". May 2, 2013, pp. 1–10. Website:
http://en.wikipedia.org/wiki/Huffman coding

Wikipedia (2013). "Entropy Encoding".
http://en.wikipedia.org/wiki/Entropy_encoding

Wikipedia (2013). "Universal Code (data compression)". May 2, 2013, pp. 1–3. Website:
http://en.wikipedia.org/wiki/Universalcoe (data compression)

Wikipedia (2013). "Data compression asymmetry". May 2, 2013, pp. 1–2. Website:
http://en.wikipedia.org/wiki/Data compression asymmetry

Wikipedia (2013). "Data Compression". May 2, 2013, pp. 1–10. Website:
http://en.wikipedia.org/wiki/Compression algorithm

Wikipedia (2013). "Polymer". May 2, 2013, pp. 1–25. Website:
http://en.wikiedia.org/wiki/Polymers

*Wikipedia (2013). "Language of mathematics", March 23, 2013, pp. 1–8. Website:
http://en.wikipedi.org/wiki/Mathematics-as-a-language

Appendices

Appendix A: A new foundation for information

Advanced Human Design

Technical Report

Volume 1, Number 1

November 2007

A New Foundation for Information

By Bradley S. Tice

Abstract: The implementation of a ternary- or quaternary-based system to information infrastructure to replace the archaic binary system. Using a ternary- or a quaternary-based system will add greater robustness, compression, and utilizability to future information systems.

Introduction: With the advent of the superior compression of both a ternary- and quaternary-based system over that of the traditional binary system in information theory, the real need for a practical application to the fundamental structure of 'information' must be re-considered for the 21st century (Tice, 2006a and 2006b). With information technology being the 'major driver' of economic growth in the past decade, adding \$2 trillion a year to the economy, the need to sustain and increase economic growth becomes an imperative (Davies, 2007: 1).

With a growing interest in 'rebuilding' the Internet, the fundamental question arises 'why be tied to an archaic binary based system when both a ternary and a quaternary based system are more robust, offer greater utilizability, and have far greater capacity for compression?' (Jesdanun, 2007: D1 and D-3 and Tice, 2006a and 2006b). The answer to this question lies with the political aspect of the innovation process. If such a system is to be built using a ternary or a quaternary based system over the out-dated binary based system, then the government must be informed of the value of such systems over the existing system of information based infrastructures (Boehlert, 2006: 1228–1229).

Part I: Information-based systems use a binary-based system represented by either a 1 or a 0. First developed by Claude Shannon in 1948 and termed 'information theory', this fundamental unit has become the 'backbone' of our information age (Gates, 1995: 30). One important aspect to information theory is that data compression, the removal of redundant features in a message, can 'reduce' the overall size of a message (Gates, 1995: 30). The need for better compression of messages is an ever growing necessity in both computing and communications (Gates, 1995: 30–31). The 2007 Nobel Prize for Physics was awarded for the discovery of GMR that has increased the capacity of computer hard drives (Cho, 2007: 179). A more profound effect to the computer industry would be the change from a binary-based system to a ternary or quaternary based system.

The Internet was a byproduct of the 'cold war'. A government sponsored project to develop a communications system that was decentralized (Nuechterlein and Weiser, 2005: 129). Under the Department of Defense's Advanced Research Project Agency: DARPA, the Internet started life as ARPAnet in 1969 (Nuechterlein and Weiser, 2005: 129). Even Tim Berners-Lee, the 'father of the web', states that "The Web is far from "done"" and that it is a "jumbled state of construction" (Berners-Lee, 1999: 158).

The out-growth of a Ph.D. dissertation, Google, the search engine company, with perhaps the most extensive computing platform in existence, wants to become an information giant (Battelle, 2005: 243,

246, and 248–249). Google is in some respect a 'Money Machine' with a value of $23 billion when it first hit the stock market in 2004 and has recorded an annual profit of $3 billion in 2006 (Vise and Malseed, 2005: 259 and Durman, 2007: 4).

Part II: The advantages of a ternary and quaternary based system over a binary based system for information theory.

Group A

Binary Non-Random Sequence

[111000111000111]

Group B

Binary Random Sequence

[111001100011111]

If Group A and Group B are compressed using the first character type and the following similar character types in a sequential order that follows that first character type, a numerical value to the number of character types can be assigned from that similar sequence of characters that can be represented by a multiple of that number represented in that group. An example will be that [111] equals the character type 1 multiplied by three to equal [111]. Notice that the character type is not a numerical one and does not have a semantic value beyond being different than the other character type [0].

Using a Key Code as a index of which character is to be multiplied, and by what amount, a compressed version of the original length of characters results.

Key Code A (Group A)

$1 = x3$

$0 = x3$

Group A

Binary Non-random Sequence

[10101]

Resulting in Group A having a compression one third the original character length of 15 characters.

Key Code B (Group B)

$1 = x3$

$0 = x3$

Group B

Binary Random Sequence

Resulting in Group B having a compression of two-third the original character length of 15 characters.

If a ternary system, or radix 3 based system, was used to represent both random and non-random sequential strings, the following three character symbols can be used: [1], [0] and [Q].

Group A

A Non-random ternary sequence

[111000QQQ111000QQQ]

Total character length of 18 characters.

Group B

A Random ternary sequence

[111000QQQQ11000QQQQ]

Total character length of 18 characters.

Again use of a Key Code to compress the original total character length by use of multiplication.

Key Code A (Group A)

$1 = x3$

$0 = x3$

$Q = x3$

Group A Non-random Ternary Sequence

[10Q10Q]

Total compression for Group A is a length of 6 characters from the original 18 character length. This is one third the original character length.

Key Code B (Group B)

> 1 = x3
>
> 0 = x3
>
> Q = x4

Group B Random Ternary Sequence

> [1_0_Q_110_QJ]

Total compression for Group B is a length of 7 characters from the original 18 character length. This is less than one half of the original character length.

If a quaternary, or radix 4 based system, was used to represent both random and non-random sequential strings, the following character symbols can be used: [1], [0], [Q], and [I].

Group A

A non-random quaternary sequence

> [111000QQQIII111000QQQIII]

Total character length of 24 characters.

Group C

A random quaternary sequence

> [1110000QQQIII1111OOQQISI]

Total character length of 24 characters.

The use of a Key Code to compress the original character length by use of multiplication.

Key Code A (Group A)

> 1 = x3
>
> 0 = x3
>
> Q = x3
>
> I = x3

Group A

Non-random quaternary sequence [10QI10QI]

Total compression for Group A is a length of 8 characters from the original 24 character length. This is one third the original character length.

Key Code C (Group C)

    1 = x
    O = x
    Q = x
    l = x

Group C

Random Quaternary Sequence

    [1110_Q_M_OOQQIJ]

Total compression for Group C is a length of 12 characters from the original 24 character length. This is one half the original character length.

Part III: In 2000 the 'Millennium Bug', or Y2K problem, arose from the perceived problem of information systems changing from one century mark to another. The concern over this problem was global in scope. Imagine the entire information system of the world being made 'redundant' by a superior information system. The concern I have for the United States is that a foreign power will implement a ternary or quaternary based information system that will 'outdate' existing binary based systems. The reason for this paper is to educate policy makers to the potential power of both a ternary and quaternary based information systems (Boehlert, 2006).

References

Battelle, J. (2005). The Search. New York: Portfolio.

Berners-Lee, T. (1999). Weaving the Web. San Francisco: Harper.

Boehlert, S. (2006). "Explaining science to Power: Make it simple, make it pay". Science, Volume 314, November 24, 2006, pp. 1228–1229.

Cho, A. (2007). "Effect that revolutionized hard drives nets a nobel". Science, Volume 318, October 12, 2007, p. 179.

Davies, F. (2007). "Impact of information technology touted". Silicon Valley.com, Wednesday March 14, 2007, p. 1.

Durman, P. (2007). "Man who took google global".

Times On Line (The Sunday Times), Sunday, May 20, 2007, pp. 1–4.

Gates, B., Myhrvold, N. and Rinearson, P. (1995). The Road Ahead. New York: Viking.

Jesdanun, A. (2007). "Rebuilding the internet".

The Modesto Bee, Thursday, April 19, 2007, pp. D-1 and D-3.

Nuecherlein, J.E. and Weiser, P.J. (2005). Digital Crossroads. Cambridge: The MIT Press.

Vise, D.A. and Malseed, M. (2005). The Google Story. New York: Delacorte Press.

Appendix B: Compression and geometric data

Abstract: Kolmogorov Complexity defines a random binary sequential string as being less patterned than a non-random binary sequential string. Accordingly, the non-random binary sequential string will retain the information about its original length when compressed; whereas the random binary sequential string will not retain such information. In introducing a radix 2 based system to a sequential string of both random and non-random series of strings using a radix 2, or binary, based system. When a program is introduced to both random and non-random radix 2 based sequential strings that notes each similar subgroup of the sequential string as being a multiple of that specific character and affords a memory to that unit of information during compression, a sub-maximal measure of Kolmogorov Complexity results in the random radix 2 based sequential string. This differs from conventional knowledge of the random binary sequential string compression values.

PACS numbers: 89.70 Eg, 89.70 Hj, 89.75 Fb, 89.75 Kd

Traditional literature regarding compression values of a random binary sequential string has an equal measure to length that is not reducible from the original state [1]. Kolmogorov complexity states that a random sequential string is less patterned than a non-random sequential string and that information about the original length of the non-random string will be retained after compression [2]. Kolmogorov complexity is the result of the development of Algorithmic Information Theory that was discovered in the mid-1960s [3]. Algorithmic Information Theory is a subgroup of Information Theory that was developed by Shannon in 1948 [4].

Recent work by the author has introduced a radix 2 based system, or a binary system, to both random and non-random sequential strings [5]. A patterned system of segments in a binary sequential string as represented by a series of 1's and 0's is rather a question of perception of subgroups within the string, rather than an innate quality of the string itself. While Algorithmic Information

Theory has given a definition of patterned verses patternless in sequential strings as a measure of random verses non-random traits, the existing standard for this measure for Kolmogorov Complexity has some limits that can be redefined to form a new sub-maximal measure of Kolomogorov Complexity in sequential binary strings [6]. Traditional literature has a non-random binary sequential string as being such: [111000111000111] resulting in total character length of 15 with groups of 1's and 0's that are sub-grouped in units of threes. A random binary sequence of strings will look similar to this example: [110100111000010] resulting in a mixture of subgroups that seem 'less patterned' than the non-random sample previously given.

Compression is the quality of a string to reduce from its original length to a compressed value that still has the property of 'decompressing' to its original size without the loss of information inherent in the original state before compression.

This original information is the quantity of the strings of original length before compression, bit length, as measured by the exact duplication of the 1's and 0's found in that original sequential string. The measure of the string's randomness is just a measure of the patterned quality found in the string.

The quality of 'memory' of the original pre-compressed state of the binary sequential string has to do with the quantity of the number of 1's and 0's in that string and the exact order of those digits in the original string are the measure of the ability to compress in the first place. Traditional literature has a non-random binary sequential string as being able to compress, while a random binary sequential string will not be able to compress. But if the measure of the number and order of digits in a binary sequence of strings is the sole factor for defining a random or non-random trait to a binary sequential string, then it is possible to 'reduce' a random binary sequential string by some measure of itself in the form of sub-groups.

These sub-groups, while not being as uniform as a non-random subgroup of a binary sequential string, will nonetheless compress from the original state to one that has reduced the redundancy in the string by implementing a compression in each subgroup of the random binary sequential string. In other words, each sub-group of the random binary sequential string will compress, retain the memory of that pre-compression state, and then, when decompressed, produce the original number and order to random binary sequential string.

The memory aspect to the random binary sequential string is, in effect, the retaining of the number and order of the information found in the original pre-compression state. This can be done by assigning a relation to the subgroup that has a quality of reducing and then returning to the original state that can be done with the use of simple arithmetic. By assigning each subgroup in the random binary sequential string with a value of the multiplication of the amount found in that sub-group, a quantity is given that can be retained for use in reducing and expanding to the original size of that quantity and can be represented by a single character that represents the total number of characters found in that sub-group.

This is the very nature of compression and duplicates the process found in the non-random binary sequential strings. As an example, the random binary sequential string [1100010011011111] can be grouped into sub-groups as follows: {11}, {000}, {1}, {00}, {11}, {0}, and {111} with each sub-group bracketed into common families of like digits. An expedient method to reduce this string would be to take similar types and reduce to a single character that represented a multiple of the exact number of characters found in that sub-group. In this case taking the bracketed {11} and assign a multiple of 2 to a single character, 1, and then reduced it to a single character in the bracket that is underlined to note the placement of the compression. The compressed random binary sequential string would appear like this: [1000100101111] with the total character length of 13, exhibiting the loss of two characters due to the compression of the two similar sub-groups.

De-compression would be the removal of the underlining of each character and the replacement of the 1's characters to each of the sub-groups that would constitute 100% retention of the original character number and order to the random binary sequential string. This makes for a new measure of Kolmogorov Complexity in a random binary sequential string.

Summary

The use of a viable compression method for sequential binary strings has applied aspects to transmission and storage of geometric data. Future papers will explore practical applications to industry regarding applied aspects of compression to geometric data.

References

[1] Kotz, S. and Johnson, N.I. (1982). Encyclopedia of Statistical Sciences, John Wiley & Sons, New York.

[2] abide.

[3] Solomonoff, R.J. (1969). Inf. & Cont. 7, 1–22 & 224–254 (1964), A.N. Kolmogorov, Pro. Inf. & Trans. 1, 1–7 (1965) and G.J. Chaitin, Jour. ACM 16, 145–159 (1969).

[4] Shannon, C.E. (1948). Bell Labs. Tech. Jour. 27, 379–423 and 623–656.

[5] Tice, B.S. (2009). Aspects of Kolmogorov Complexity: The Physics of Information, River Publishers, Denmark.

[6] Kotz, S. and Johnson, N.I. (1982). Encyclopedia of Statistical Sciences, John Wiley & Sons, New York.

Appendix C: The analysis of binary, ternary and quaternary based systems for communications theory

Abstract: The implementation of a ternary or quaternary based system to information infrastructure to replace the archaic binary system. Using a ternary or a quaternary based system will add greater robustness, compression, and utilizability to future information systems.

Key words: Radix 2, Radix 3, Radix 4, Binary, Ternary, Quaternary, Information Theory, Communication Theory

Introduction: With the advent of the superior compression of both a ternary and quaternary based system over that of the traditional binary system in information theory, the real need for a practical application to the fundamental structure of 'information' must be re-considered for the 21 century. With information technology being the 'major driver' of economic growth in the past decade, adding $2 trillion a year to the economy, the need to sustain and increase economic growth becomes an imperative [1].

With a growing interest in 'rebuilding' the Internet, the fundamental question arises 'why be tied to an archaic binary based system when both a ternary and a quaternary based system are more robust, offer greater utilizability, and have far greater capacity for compression?' [2–4]. The answer to this question lies with the political aspect of the innovation process. If such a system is to be built using a ternary or a quaternary based system over the out-dated binary based system, then the government must be informed of the value of such systems over the existing system of information based infrastructures [5].

Part I: Information based systems use a binary based system represented by either a 1 or a 0. First developed by Claude Shannon in 1948 and termed 'information theory', this fundamental unit has become the 'backbone' of our information age [6]. One important aspect to information theory is that data compression, the removal of redundant features in a message, can 'reduce' the overall size of a message [7]. The need for better compression of messages is an ever growing necessity in both computing and communications [8]. The 2007 Nobel Prize for Physics was awarded for the discovery of GMR that has increased the capacity of computer hard drives [9]. A more profound effect to the computer industry would be the change from a binary based system to a ternary or quaternary based system.

The internet was a byproduct of the 'cold war'. A government sponsored project to develop a communications system that was decentralized [10]. Under the Department of Defense's Advanced Research Project Agency: DARPA, the internet started life as ARPAnet in 1969 [11]. Even Tim Berners-Lee, the 'father of the web', states that "The Web is far from "done"" and that it is a "jumbled state of construction" [12].

The out-growth of a Ph.D. dissertation, Google, the search engine company, with perhaps the most extensive computing platform in existence, wants to become an information giant [13]. Google is in some respect a 'Money Machine' with a value of $23 billion when it first hit the stock market in 2004 and has recorded an annual profit of $3 billion in 2006 [14 and 15].

Part II: The advantages of a ternary and quaternary based system over a binary based system for information theory.

2.1   Radix 2 Base

Radix 2 Based System

    Group A

    Binary Non-random Sequence

        [111000111000111]

    Group B

    Binary Random Sequence

        [111001100011111]

If Group A and Group B are compressed using the first character type and the following similar character types in a sequential order that follows that first character type, a numerical value to the number of character types can be assigned from that similar sequence of characters that can be represented by a multiple of that number represented in that group. An example will be that [111] equals the character type 1 multiplied by three to equal [111]. Notice that the character type is not a numerical one and does not have a semantic value beyond being different than the other character type [0].

Using a Key Code as an index of which character is to be multiplied, and by what amount, a compressed version of the original length of characters results.

Key Code A (Group A)

> 1 = x3
>
> 0 = x3
>
> Group A

Binary Non-random Sequence

> [10101]

Resulting in Group A having a compression one third the original character length of 15 characters.

Key Code B (Group B)

> 1 = x3
>
> 0 = x3

Group B

Binary Random Sequence

> [1_00110_11111]

Resulting in Group B having a compression of two-third the original character length of 15 characters.

2.2   A Radix 3 Base

Radix 3 Based System

If a ternary system, or radix 3 based system, was used to represent both random and non-random sequential strings, the following three character symbols can be used: [1], [0] and [Q].

Group A

A non-random ternary sequence

[111000QQQ111000QQQ]

Total character length of 18 characters.

Group C

A random ternary sequence

[111000QQQQ11000QQQQ]

Total character length of 18 characters.

Again use of a Key Code to compress the original total character length by use of multiplication.

Key Code A (Group A)

1 = x3

0 = x3

Q = x3

Group A

Non-random ternary sequence

[10Q10Q]

Total compression for Group A is a length of 6 characters from the original 18 character length. This is one third the original character length.

Key Code C (Group C]

1 = x3

0 = x3

Q = x4

Group C Random Ternary Sequence

[1_0_Q_110_QJ]

Total compression for Group C is a length of 7 characters from the original 18 character length. This is less than one half of the original character length.

2.3   Radix 4 Base

Radix 4 Based System

If a quaternary, or radix 4 based system, was used to represent both random and non-random sequential strings, the following character symbols can be used: [1], [0], [Q], and [I].

Group A

A non-random quaternary sequence.

[111000QQQII11110OOQQQIII]

Total character length of 24 characters.

Group D

A random quaternary sequence.

[1110000QQQIII11110OQQQIII]

Total character length of 24 characters.

The use of a Key Code to compress the original character length by use of multiplication.

Key Code A (Group A)

    1 = x3
    0 = x3
    Q = x3
    I = x3

Group A

Non-random quaternary sequence

[10QI10QI]

Total compression for Group A is a length of 8 characters from the original 24 character length. This is one third of the original character length.

Key Code D (Group D)

> 1 = x
>
> O = x
>
> Q = x
>
> l = x

Group D

Random Quaternary Sequence

> [1110_QJ_1_OOQQIJ]

Total compression for Group D is a length of 12 characters from the original 24 character length. This is one half the original character length.

Part III

In 2000 the 'Millennium Bug', or Y2K problem, arose from the perceived problem of information systems changing from one century mark to another. The concern over this problem was global in scope. Imagine the entire information system of the world being made 'redundant' by a superior information system? The concern I have for the United States is that a foreign power will implement a ternary or quaternary based information system that will 'outdate' existing binary based systems. The reason for this paper is to educate policy makers to the potential power of both a ternary and quaternary based information systems [16].

Summary: The results of using a compression engine to compress both random and non-random sequential strings of radix 2, radix 3 and radix 4 based strings resulted in the following:

Radix Base	Random Base	Non-random
Radix 2 15 character length total	11	5
Radix 3 18 character length total	7	6
Radix 4 24 character length total	12	8

Both the radix 3 and radix 4 based systems had substantial compression values in the random sequential strings categories. As random sequential strings have the most applicable nature to practical modes of information transmission and storage, these findings have both theoretical and applied aspects to communication theory in all of its manifestations.

References

1. Shannon, C.E. (1948). "A mathematical theory of communication", *Bell Sys. Tech. Jour. 27,* 379–423 and 623–656.

2. Tice, B.S. (2006). "A radix 4 based system for communications theory", Technical Paper, *Advanced Human Design,* Volume 1, Number 2, 1–3.

3. Li, M. and Vitanyi, P.M.H. (1993/1997). An Introduction to Kolmogorov Complexity and its Applications. Springer, New York.

4. Martin-Lof, P. (1966). "The definition of random sequences". *Information and Control.* 9, 602–619 (1966).

5. Tice, Ibide.

Appendix D: The use of a radix 5 base for transmission and storage of information

Abstract: The radix 5 based system employs five separate characters that have no semantic meaning except not representing the other characters. Traditional literature has a random string of binary sequential characters as being "less patterned" than non-random sequential strings. A non-random string of characters will be able to compress, where as a random string of characters will not be able to compress. This study has found that a radix 5 based character length allows for equal compression of random and non-random sequential strings. This has important aspects to information transmission and storage.

Key words: Radix 5, Information Theory, Algorithmic Information Theory, transmission, storage, Communication Theory.

Introduction: As communications handle an ever-growing amount of information for transmission and storage, the very real need for an upgrade in the fundamental structure of such a system has come to light. As the very bases of coding is compression, the greater the amount of information compressed, the more efficient the system. The earliest calculating machine was the human hand, its five digits representing a natural symmetry found, with frequency, in the organic world [1 and 2]. A radix 5 base system, also known as a quinary numeral system, is composed of five separate characters that have no meaning apart from the fact the each character is different than the other characters. This is a development from the binary system used in Shannon's information theory (1948) [3].

Part I: The radix 5 base is not the traditional binary based error-detection and error-correcting codes that are also known as 'prefix codes' that use a 5-bit length for decimal coding [4]. A radix 5 base is composed of five separate symbols with each an individual character with no semantic meaning. A random string of symbols has the quality of being 'less patterned' than a non-random string of symbols. Traditional literature on the subject of compression, the ability for a string to reduce in size while retaining 'information' about its original character size, states that a non-random string of characters will be able to compress, where as the random string of characters will not compress [5].

Part II: The following examples will use the following symbols for a radix 5 based system of characters [Example A]

Example A

o

O

Q

1

I

The following is an example of compression of a random and non-random radix 5 base system. A non-random string of radix 5 based characters with a total 15 character length [Group A].

Group A oooOOOQQQ111III

A random string of a radix 5 based characters with a total of 15 character length [Group B].

Group B oooOOQQQQ11IIIII

If a compression program were to be used on Group A and Group B that consisted of underlining the first individual character of a similar group of sequential characters, moving towards the right, on the string and multiplying it by a formalized system of arithmetic as found in a key, see Key Code A and Key Code B, with the compression of Group A and Group B as the final result.

Key Code A (For Group A)

$o = x3$

$O = x3$

$Q = x3$

$I = x3$

$1 = x3$

Group A   o O Q 1 I

Resulting in a 5 character length for Group A.

Key Code (For Group B)

o = x3

O = x2

Q = x4

1 = x2

I – x4

Group B oOQ1I

Resulting in a 5 character length for Group B.

Both Group A (Non-random) and Group B (Random) have the same compression values, each group resulted in a compression value of 1/3 the total pre-compression, original, state. This contrasts traditional notions of random and non-random strings [6]. These findings are similar to Tice (2003) and have applications to both Algorithmic Information Theory and Information Theory [7].

Some other examples using Example A Radix 5 characters [oOQ1I] to test random and non-random sequential strings.

The following is a non-random string of a radix 5 based characters with a total of 15 character length [Group A].

Group A oooOOOQQQ 111 III

A random string of a radix 5 based characters with a total of 15 character length (Group C).

Group C oooooOQQQQQ1I

If a compression program were to be used on group A and group C that consist of underlining the first individual character of a similar group of sequential characters, moving towards the right, on the string and multiplying it by a formalized system of arithmetic as found in a key, see Key Code A and Key Code C, with the compression of group A and group C as the final result.

Key Code C (For Group A)

o = x3

O = x3

Q = x3

1 = x3

I = x3

Group A oOQ1I

Resulting in a 5 character length for Group A.

Key Code C (For Group C)

o = x5

O = xl

Q = x5

1 = xl

I = x1

Group C oOQ1I

Resulting in a 5 character length for Group C.

This example has Group A as a non-random string and Group D as a random string using radix 5 characters for a total 15 character length.

A non-random string of radix 5 characters with a 15 character length (Group A). Group A oooOOOQQQ 111 III

A random string of a radix 5 based characters with a total of 15 character length (Group D).

Group D oOOOOQQ 1111 IIII

If a compression program were to be used on Group A and Group D that consisted of underlining the first individual character of a similar group of sequential characters, moving towards the right, on the string and multiplying it by a formalized system of arithmetic as found in a key, see Key Code A and Key Code D, with the compression of Group A and Group D as the final result.

Key Code A (For Group A)

O = x3

O = x3

Q = x3

1 = x3

I = x3

Group A oOQ1I

Resulting in a 5 character length for Group A.

Key Code D (For Group D)

o = xl

O = x4

Q = xl

1 = x4

I = x4

Group D oOQ1I

Resulting in a 5 character length for Group D.

As a final example, Group A is a non-random sequential string and Group E is a random sequential string using a radix 5 characters for a total of 15 character length.

A non-random string of radix 5 based characters with a 15 character length (Group A). Group A oooOOOQQQ 111 III

A random string of a radix 5 based characters with a total of 15 character length (Group E). Group E ooOOOOQQQQ 111II

If a compression program were to be used on Group A and Group E that consisted of underlining the first individual character of a similar group of sequential characters, moving to the right, on the string and multiplying it by a formalized system of arithmetic as found in a key, see Key Code A and Key Code E, with the compression of Group A and Group E as the final result.

Key Code A (For Group A)

o = x3

O = x3

Q = x3

1 = x3

I = x3

Group A oOQ1I

Resulting in a 5 character length for Group A.

Key Code E (For Group E)

o = x2

O = x4

Q = x4

1 = x3

I = x2

Group E oOQ1I

Resulting in a 5 character length for Group E.

Again, these examples conflict with traditional notions of random and non-random sequential strings in that the compression ratio is one third that of the original character number length for both the random and non-random sequential strings using a radix 5 base system.

Part III: Traditional information based systems use a binary-based system represented by either a 1 or a 0. First developed by Claude Shannon in 1948 and termed 'information theory', this fundamental unit has become the backbone of our information age. One important aspect to information theory is that of data compression, the removal of redundant features in a message that can reduce the overall size of a message [8]. With the substantial compression values found in using a radix 5 based system, it seems a new paradigm has arrived to carry the future of information.

Information technology has been the major driver of the economic growth in the past decade adding $2 trillion a year to the economy [9]. This growth needs to be sustained in order for new jobs and the economy to maintain a high standard of living. Only by considering alternative developments to existing models of technology, can the future of the economy develop and continue at a successful level of growth.

The internet was an outgrowth of the cold war as a government-sponsored project to develop a communications network that was decentralized [10]. Today the internet is the major highway of global information with search engine technology rapidly taking center stage on both universities research departments as well as the Dow Jones index. The need to handle this vast and ever growing amount of information will need a fundamental change to the very nature of the structure of our information systems. It is clear that any new developments to deal with more and more information must begin at the fundamental level.

With a radix 5 base that has been proved to have the compression ratio similar in both random and non-random states, the question of usage as a medium for transmission and storage of information becomes paramount. With an ever increasing need for transmission and storage in the areas of telecommunications and computer science, the viability of a new system at the fundamental level of communication theory that is both robust and diverse enough to allow for future growth beyond the binary based system in use today.

Summary: This paper has shown that a radix 5 based system has profound properties of compression that are well beyond those found in binary systems using sequential strings of a random and non-random types. These compression values have strong potential applications to information theory and communication theory as a whole.

While the identical compression values for random and non-random radix 5 based strings is a result of this paper, the application of this theory to communication theory cannot be understated. It has been shown that a radix 5 based system has a compression factor that makes it an ideal functional standard for future information systems, particularly in the fields of telecommunications and computer science.

References:

1. Ifrah, G. (2000). *The Universal History of Numbers,* John Wiley & Sons, Inc., New York, p. 47.

2. Weyl, H., "Symmetry", p. 710, In Newman, J. R. (Editor). *The World of Mathematics,* Simon and Schuster, New York, pp. 671–724.

3. Shannon, C.E. (1948). "A mathematical theory of communication", *Bell Sys. Tech. Jour.* 21, pp. 2–19 and 623–656.

4. Richards, R.K. (1955). *Arithmetic Operations in Digital Computers,* D. Van Nostrand Company, Inc., Princeton, p. 184.

5. Kotz, S. and Johnson, N.I. (1982). *Encyclopedia of Statistical Sciences,* John Wiley & Sons, Inc., NewYork, p. 39.

6. Kotz, S. and Johnson, N.L. (1982). *Encyclopedia of Statistical Sciences,* John Wiley & Sons, Inc., New York, p. 39.

7. Tice, B.S. (2003). *Two Models of Information,* 1st Books Publishers, Bloomington.

8. Gates, B., Myhrvold, N. and Rinearson, P. *The Road Ahead,* Viking, New York, p. 30.

9. Davis, F. (2007). "Impact of information technology touted", *Silicon Valley.com,* p. 1.

10. Nuecherlein J.E. and Weiser, P.J. (2005). *Digital Crossroads,* The MIT Press, Cambridge, p. 129.

Appendlx E: A comparison of a radix 2 and a radix 5 based systems

Abstract: A radix 2 based system is composed of two separate character types that have no meaning except not representing the other character type as defined by Shannon in 1948. The radix 5 based system employs five separate characters that have no semantic meaning except not representing the other characters. Traditional literature has a random string of binary sequential characters as being "less patterned" than non-random sequential strings. A non-random string of characters will be able to compress, where as a random string of characters will not be able to compress. This study has found that a radix 5 based character length allows for equal compression of random and non-random sequential strings. This has important aspects to information transmission and storage.

Key words: Radix 5, Information Theory, Algorithmic Information Theory, Transmission, Storage, Communication Theory.

Introduction: As communications handle an ever-growing amount of information for transmission and storage, the very real need for an upgrade in the fundamental structure of such a system has come to light. As the very bases of coding is compression, the greater the amount of information compressed, the more efficient the system. The earliest calculating machine was the human hand, its five digits representing a natural symmetry found, with frequency, in the organic world [1 and 2]. A radix 5 base system, also known as a quinary numeral system, is composed of five separate characters that have no meaning apart from the fact the each character is different than the other characters. This is a development from the binary system used in Shannon's information theory (1948) [3].

The radix 2 based system: The radix 2 based system is a two character system that has no semantic meaning except not representing the other character type. The traditional 1 and 0 will be used in this paper.

The following is an example of compression of a random and non-random radix 2 based system. A non-random sequential string of characters will have a total length of 15 characters as seen in Group A.

Group A: 111000111000111

A random sequential string of characters will have a total length of 15 characters as seen in Group B.

Group B: 110000111110111

If a compression program were to be used on Group A and Group B that consisted of underlining the first individual character of a similar group of sequential characters, moving towards the right, on the string and multiplying it by a formalized system of arithmetic as found in a key, see Key Code 1 and Key Code 2, with the compression of Group A and Group B as a final result.

Key Code 1 (For Group A)

I = x3

0 = x3

Group A 10101

Resulting in a compressed state of 5 characters for Group A.

Key Code 2 (For Group B)

I = x5

0 = x4

Group B 11010111

Resulting in a compressed state of 8 characters for Group B.

Compression values of the non-random binary sequential string are one third the original 15 character length and the random binary sequential string are almost half of the original random 15 character length.

Part I: The radix 5 base is not the traditional binary based error-detection and error-correcting codes that are also known as 'prefix codes' that use a 5-bit length for decimal coding [4]. A radix 5 base is composed of five separate symbols with each an individual character with no semantic meaning. A random string of symbols has the quality of being 'less patterned' than a non-random string of symbols. Traditional literature on the subject of compression, the ability for a string to reduce in size while retaining 'information' about its original character size, states that a non-random string of characters will be able to compress, where as the random string of characters will not compress [5].

Part II: The following examples will use the following symbols for a radix 5 based system of characters [Example A]

Example A

o

O

Q

1

I

The following is an example of compression of a random and non-random radix 5 base system. A non-random string of radix 5 based characters with a total 15 character length [Group A].

Group A oooOOOQQQ111III

A random string of a radix 5 based characters with a total of 15 character length [Group B].

Group B oooOOQQQQ11IIIII

If a compression program were to be used on Group A and Group B that consisted of underlining the first individual character of a similar group of sequential characters, moving towards the right, on the string and multiplying it by a formalized system of arithmetic as found in a key, see Key Code A and Key Code B, with the compression of Group A and Group B as the final result.

Key Code A (For Group A)

o = x3

O = x3

Q = x3

1 = x3

I = x3

Group A o O Q 1 I

Resulting in a 5 character length for Group A.

Key Code B (For Group B)

o = x3

O = x2

Q = x4

1 = x2

I = x 4

Group B o O Q 1 I

Resulting in a 5 character length for Group B.

Both Group A (Non-random) and Group B (Random) have the same compression values, each group resulted in a compression value of 1/3 the total pre-compression, original, state. This contrasts traditional notions of random and non-random strings [6]. These findings are similar to Tice (2003) and have applications to both Algorithmic Information Theory and Information Theory [7].

Some other examples using Example A Radix 5 characters [oOQII] to test random and non-random sequential strings.

The following is a non-random string of a radix 5 based characters with a total of 15 character length [Group A].

Group A oooOOOQQQ111III

A random string of a radix 5 based characters with a total of 15 character length (Group C).

Group C oooooOQQQQQ1I

If a compression program were to be used on group A and group C that consist of underlining the first individual character of a similar group of sequential characters, moving towards the right, on the string and multiplying it by a formalized system of arithmetic as found in a key, see Key Code A and Key Code C, with the compression of group A and group C as the final result.

Key Code A (For Group A)

o = x3

O = x3

Q = x3

1 = x3

1 = x3

Group A oOQ11

Resulting in a 5 character length for Group A.

Key Code C (For Group Q)

o = x5

O = xl

Q = x5

1 = xl

I = xl

Group C oOQ1I

Resulting in a 5 character length for Group C.

This example has Group A as a non-random string and Group D as a random string using radix 5 characters for a total 15 character length.

A non-random string of radix 5 characters with a 15 character length (Group A). Group A oooOOOQQQ 111 III

A random string of a radix 5 based characters with a total of 15 character length (Group D). Group D oOOOOQQ1111IIII

If a compression program were to be used on group A and group D that consisted of underlining the first individual character of a similar group of sequential characters, moving towards the right, on the string and multiplying it by a formalized system of arithmetic as found in a key, see Key Code A and Key Code D, with the compression of group A and group D as the final result.

Key Code A (For Group A)

o = x3

O = x3

Q = x3

1 = x3

I = x3

Group A oOQII

Resulting in a 5 character length for Group A.

Key Code D (For Group D)

o = xl

O = x4

Q = xl

1 = x4

I = x4

Group D oOQ1I

Resulting in a 5 character length for Group D.

As a final example Group A is a non-random sequential string and Group E as a random sequential string using a radix 5 characters for a total of 15 character length.

A non-random string of radix 5 based characters with a 15 character length (Group A). Group A oooOOOQQQ 111 III

A random string of a radix 5 based characters with a total of 15 character length (Group E). Group E ooOOOOOQQQQ 11111

If a compression program were to be used on group A and group E that consisted of underlining the first individual character of a similar group of sequential characters, moving to the right, on the string and multiplying it by a formalized system of arithmetic as found in a key, see Key Code A and Key Code E, with the compression of group A and group E as the final result.

Key Code A (For Group A)

o = x3

O – x3

Q = x3

1 = x3

I = x3

Group A oOQ1I

Resulting in a 5 character length for Group A.

Key Code E (For Group E).

o = x2

O = x4

Q = x4

1 = x3

I = x2

Group E oOQ1I

Resulting in a 5 character length for group E.

Again, these examples conflict with traditional notions of random and non-random sequential strings in that the compression ratio is one third that of the original character number length for both the random and non-random sequential strings using a radix 5 base system.

Part III: Traditional information-based systems use a binary-based system represented by either a 1 or a 0. First developed by Claude Shannon in 1948 and termed 'information theory', this fundamental unit has become the backbone of our information age. One important aspect to information theory is that of data compression, the removal of redundant features in a message that can reduce the overall size of a message [8]. With the substantial compression values found in using a radix 5 based system, it seems a new paradigm has arrived to carry the future of information.

Information technology has been the major driver of the economic growth in the past decade adding $2 trillion a year to the economy [9]. This growth needs to be sustained in order for new jobs and the economy to maintain a high standard of living. Only by considering alternative developments to existing models of technology, can the future of the economy develop and continue at a successful level of growth.

The internet was an outgrowth of the cold war as a government-sponsored project to develop a communications network that was decentralized [10]. Today the internet is the major highway of global information with search engine technology rapidly taking center stage on both universities research departments as well as the Dow Jones index. The need to handle this vast and ever growing amount of information will need a fundamental change to the very nature of the structure of our information systems. It is clear that any new developments to deal with more and more information must begin at the fundamental level.

With a radix 5 base that has been proved to have the compression ratio similar in both random and non-random states, the question of usage as a medium for transmission and storage of information becomes paramount. With an ever increasing need for transmission and storage in the areas of telecommunications and computer science, the viability of a new system at the fundamental level of communication theory that is both robust and diverse enough to allow for future growth beyond the binary based system in use today.

Summary: This paper has shown that a radix 5 based system has profound properties of compression that are well beyond those found in binary systems using sequential strings of a random and non-random types. These compression values have strong potential applications to information theory and communication theory as a whole.

When comparing the radix 2 and the radix 5 based systems, the greater compression factor of the radix 5 based system has strong applications to signal transmission and storage issues.

While the identical compression values for random and non-random radix 5 based strings is a result of this paper, the application of this theory to communication theory cannot be understated. It has been shown that a radix 5 based system has a compression factor that makes it an ideal functional standard for future information systems, particularly in the fields of telecommunications and computer science.

References

1. Ifrah, G. (2000). *The Universal History of Numbers,* John Wiley & Sons, Inc., New York, p. 47.

2. Weyl, H. (1956). "Symmetry", p. 710, In: Newman, J. R. (Editor) *The World of Mathematics,* Simon and Schuster, New York, pp. 671–724.

3. Shannon, C.E. (1948). "A mathematical theory of communication", *Bell Sys. Tech. Jour21,* pp. 379–423and 623–656.

4. Richards, R.K. (1955). *Arithmetic Operations in Digital Computers,* D. Van Nostrand Company, Inc., Princeton, p. 184.

5. Kotz, S. and Johnson, N.I. (1982). *Encyclopedia of Statistical Sciences,* John Wiley & Sons, Inc., NewYork, p. 39.

6. Tice, B.S. (2003). *Two Models of Information,* 1st Books Publishers, Bloomington.

7. Gates, B., Myhrvold, N. and Rinearson, P. (1995). *The Road Ahead,* Viking, New York, p. 30.

8. Davis, F. (2007). "Impact of information technology touted", *Silicon Valley.com,* March 14, 2007, p. 1.

9. Nuecherlein J.E. and Weiser, P.J. (2005). *Digital Crossroads,* The MIT Press, Cambridge,

10. p. 129.

Appendix F: Random and non-random sequential strings using a radix 5 based system

Abstract: Kolmogorov Complexity defines a random binary sequential string as being less patterned than a non-random binary sequential string. Accordingly, the non-random binary sequential string will retain the information about its original length when compressed, where as the random binary sequential string will not retain such information. In introducing a radix 5 based system to a sequential string of both random and non-random series of strings using a radix 5, or quinary, based system. When a program is introduced to both random and non-random radix 5 based sequential strings that notes each similar subgroup of the sequential string as being a multiple of that specific character and affords a memory to that unit of information during compression, a sub-maximal measure of Kolmogorov Complexity results in the random radix 5 based sequential string. This differs from conventional knowledge of the random binary sequential string compression values.

PACS numbers: 89.70 Eg, 89.70 Hj, 89.75 Fb, 89.75 Kd

Traditional literature regarding compression values of a random binary sequential string have an equal measure to length that is not reducible from the original state [1]. Kolmogorov complexity states that a random sequential string is less patterned than a non-random sequential string and that information about the original length of the non-random string will be retained after compression [2]. Kolmogorov complexity is the result of the development of Algorithmic Information Theory that was discovered in the mid-1960s [3]. Algorithmic Information Theory is a subgroup of Information Theory that was developed by Shannon in 1948 [4].

Recent work by the author has introduced a radix 5 based system, or a quinary system, to both random and non-random sequential strings [5]. A patterned system of segments in a binary sequential string as represented by a series of 1's and 0's is rather a question of perception of subgroups within the string, rather than an innate quality of the string itself. While Algorithmic Information Theory has given a definition of patterned verses patternless in sequential strings as a measure of random verses non-random traits, the existing standard for this measure for Kolmogorov Complexity has some limits that can be redefined to form a new sub-maximal measure of Kolomogorov Complexity in sequential binary strings [6]. Traditional literature has a non-random binary sequential string as being such: [111000111000111] resulting in total character length of 15 with groups of 1's and 0's that are sub-grouped in units of threes. A random binary sequence of strings will look similar to this example: [110100111000010] resulting in a mixture of subgroups that seem 'less patterned' than the non-random sample previously given.

Compression is the quality of a string to reduce from its original length to a compressed value that still has the property of 'decompressing' to its original size without the loss of the information inherent in the original state before compression. This original information is the quantity of the strings original length before compression, bit length, as measured by the exact duplication of the 1's and 0's found in that original sequential string. The measure of the string's randomness is just a measure of the patterned quality found in the string.

The quality of 'memory' of the original pre-compressed state of the binary sequential string has to do with the quantity of the number of 1's and 0's in that string and the exact order of those digits in the original string are the measure of the ability to compress in the first place. Traditional literature has a non-random binary sequential string as being able to compress, while a random binary sequential string will not be able to compress. But if the measure of the number and order of digits in a binary sequence of strings is the sole factor for defining a random or non-random trait to a binary sequential string, then it is possible to 'reduce' a random binary sequential string by some measure of itself in the form of sub-groups. These sub-groups, while not being as uniform as a non-random sub-group of a binary sequential string, will nonetheless compress from the original state to one that has reduced the redundancy in the string by implementing a compression in each subgroup of the random binary sequential string. In other words, each sub-group of the random binary sequential string will compress, retain the memory of that pre-compression state, and then, when decompressed, produce the original number and order to random binary sequential string.

The memory aspect to the random binary sequential string is, in effect, the retaining of the number and order of the information found in the original pre-compression state. This can be done by assigning a relation to the subgroup that has a quality of reducing and then returning to the original state that can be done with the use of simple arithmetic. By assigning each subgroup in the random binary sequential string with a value of the multiplication of the amount found in that sub-group, a quantity is given that can be retained for use in reducing and expanding to the original size of that quantity and can be represented by a single character that represents the total number of characters found in that sub-group. This is the very nature of compression and duplicates the process found in the non-random binary sequential strings. As an example, the random binary sequential string [110001001101111] can be grouped into sub-groups as follows: {11}, {000}, {1}, {00}, {11}, {0}, and {111} with each sub-group bracketed into common families of like digits. An expedient method to reduce this string would be to take similar types and reduce to a single character that represented a multiple of the exact number of characters found in that sub-group. In this case taking the bracketed {11} and assign a multiple of 2 to a single character, 1, and then reduced it to a single character in the bracket that is underlined to note the placement of the compression.

The compressed random binary sequential string would appear like this: [1000100101111] with the total character length of 13, exhibiting the loss of two characters due to the compression of the two similar sub-groups. De-compression would be the removal of the underlining of each character and the replacement of the 1's characters to each of the sub-groups that would constitute a 100% retention of the original character number and order to the random binary sequential string. This makes for a new measure of Kolmogorov Complexity in a random binary sequential string.

This same method of compression can be used with a radix 5 based system that provides for an even greater measure of reduction than is found in the binary sequential string. The radix 5 base number system has five separate characters that have no semantic meaning except not representing the other characters in the five character system. The following five numbers will represent the five characters found in the radix 5 base number system that will be used as an example in this paper: [0, 1, 2, 3 & 4]. As an example of a random radix 5 sequential string the following would appear like this: [001112233334440111223444] with a total character length of 24 characters. If all the applicable similar sequential 3 characters are compressed to a single representative character that represents the other two characters in the three character compressed unit of the string, then the following would result: [0012233334012234].

The underlined characters represent the initial position of the three character group of similar characters with a compressed state of 16 characters total. This is a reduction of one third the total original character length of 24 characters. A non-random radix 5 base sequential string will have the same character types: [0, 1, 2, 3 & 4] but with a regular pattern of groupings such as [00112233440011223344] that has a total character length of 20 and if all two sequentially similar characters are compressed using all 5 character types the following will occur: [0123401234] resulting in a compressed non-random radix 5 base sequential string of 10.

The paper has shown that a sub-maximal measure of Kolmogorov complexity exists that has implications to a new standard of the precise measure of randomness in both a radix 2 and a radix 5 based number systems.

References

[1] S. Kotz and N.I. Johnson, Encyclopedia of Statistical Sciences (John Wiley & Sons, New York, 1982).

[2] abide.

[3] R.J. Solomonoff, Inf. & Cont. 7, 1–22 & 224–254 (1964), A.N. Kolmogorov, Pro. Inf. & Trans. 1, 1–7 (1965) and G.J. Chaitin, Jour. ACM 16, 145–159 (1969).

[4] C.E. Shannon, Bell Labs. Tech. Jour. 27, 379–423 and 623–656 (1948).

[5] B.S. Tice "The use of a radix 5 base for transmission and storage of information", Poster for the Photonics West Conference, San Jose, California Wednesday January 23, 2008.

[6] S. Kotz and N.I. Johnson, Encyclopedia of Statistical Sciences (John Wiley & Sons, New York, 1982).

Appendix G: A comparison of compression values of binary and ternary base systems

Abstract: The paper introduces the ternary, or radix 3, based system for use as a fundamental standard beyond the traditional binary, or radix 2, based system in use today. A compression level is noted that is greater than the known Martin-L of standard of randomness in both binary and ternary sequential strings.

Key words: Radix 2, Binary, Radix 3, Ternary, Information Theory, Compression Ratio

I. Introduction: A ternary, or radix 3, based system is defined as three separate characters, or symbols, that have no semantic meaning apart from not representing the other characters. This is the same notion Shannon gave to the binary based system used in his paper's on information theory upon its publication in 1948 (Shannon, 1948). Richards has noted that the radix 3 based system as the most efficient base, more so than even the radix 2 or radix 4 based systems (Richards, 1955: 8–9). A compression level is noted in this paper that is greater than the known Martin-L of standard of randomness in both binary and ternary sequential strings.

II. Randomness: The earliest definition for randomness in a string of 1's and 0's was defined by von Mises, but it was Martin-L of's paper of 1966 that gave a measure to randomness by the patternless of a sequence of 1's and 0's in a string that could be used to define a random binary sequence in a string (Martin-L of, 1966). This is the classical measure for Kolmogorov complexity, also known as Algorithmic Information Theory, of the randomness of a sequence found in a binary string (Kotz and Johnson, 1982: 39). Martin-L of (1966) also defined a random binary sequential string as being unable to compress from its original state. Nonrandom binary sequential strings can compress to less than their original state (Martin-L of, 1966). Paper accepted and prepared for poster session Wednesday September 3, 2008 for the Royal Statistical Society 2008 Conference in Nottingham, England, United Kingdom, September 1 5, 2008.

III. Compression Program: The compression program to be used has been termed the Modified Symbolic Space Multiplier Program as it simply notes the first character in a line of characters in a binary sequence of a string and subgroups them into common or like groups of similar characters, all 1's grouped with 1's and all 0's grouped with 0's, in that string and is assigned a single character notation that represents the number found in that sub-group, so that it can be reduced, compressed, and decompressed, expanded, back to its original length and form. An underlined 1 or 0 is usually used to note the notation symbol for the placement and character type in previous applications of this program. An italicized character will be used for this paper.

IV. Binary System: The binary system, also known as a radix 2 based system, is composed of two characters, usually a 0 and a 1, that have no semantic properties except not representing the other. Group A will represent a nonrandom sequential binary string and Group B will represent a random sequential binary string. Both Group A and Group B will be 15 characters in total length.

Group A: [000111000111000] (Nonrandom)

Group B: [001110110011100] (Random)

Utilizing the Modified Symbolic Space Multiplier Program to process like sequential characters, either 0's or 1's, into sub-groups and note them with an italicized character specific to that sub-group and having it represent a specific multiple of that sub-group as found in a key, in this case Group A Key and Group B Key, as a compressed aspect to both Group A and Group B sequential binary strings.

Group A Key: All italicized characters will represent a multiple of 3.

Group B Key: The italicized character 0 will represent a multiple of 2 and the italicized character 1 will represent a multiple of 3.

Group A: [01010]

Group B: [02011010]

The compressed state of Group A, nonrandom, is five characters in length. The compressed state of Group B, random, is 8 characters in length. Note that the random sequential binary string in Group B compressed to less than the original total pre-compression length. This differs from standards known in Martin-L of randomness and those found in Kolmogorov Complexity.

V. Ternary System: In a ternary, or radix 3, based system, there are three characters used that have no semantic meaning except not representing the other two characters. Group C will represent a nonrandom ternary sequential string and Group D will represent a random ternary sequential string. The total length for each group, Group C and Group D, will be 12 characters in length. The three characters to be used in this study are 0, 1, and 2.

Group C: [001122001122]

Group D: [001222011222]

Again each group will be assigned a specific compression multiple based on a specific character type, in this case an italicized 0, 1, and 2, as defined in a key, Group C Key and Group D Key.

Group C Key: The italicized characters 0, 1 and 2 will represent each a multiple of 2.

Group D Key: The italicized character 0 will represent a multiple of 2. The italicized character 1 will represent a multiple of 2 and the italicized character 2 will represent a multiple of 3.

Group C: {012012}

Group D: [012012]

The compressed state of group C, nonrandom, is 6 characters in length. The compressed string of Group D, random, is 6 characters in length. Again note that Group D, the random sequential ternary string, is less than its pre-compressed state, and again, is novel for those extrapolations of binary examples found in Kolmogorov Complexity.

VI. Application of Theory: Compression of data for transmission and storage of information is the most practical application of a binary system in telecommunications and computing. The application of a ternary, or radix 3, based system to existing communication systems has many advantages. The first is the greater amount of compression from this base, as opposed to the standard binary based system in use today, of random strings of data, and secondly, as a more utilizable system because of the three character, or symbol, based system that provides for more variety to develop information applications. From telecommunications to computing, the ternary-based system applied at a fundamental standard would allow for a more robust communications system than is currently used today.

References

Shannon, C.E. (1948). A mathematical theory of communication, *Bell Sys. Tech. Jour.* 27, pp. 379–423 and 623–656.

Richards, R.K. (1955). *Arithmetic Operations in Digital Conguters,* New York: Van Nostrand Company, Inc.

Martin-L of, P. (1966). The definition of random sequences, *Info. & Cont.,* 9, 602–619.

Kotz, S. and Johnson, N.L. (1982). *Encyclopedia of Statistical Sciences,* New York: John Wiley & Sons.

Appendix H: Patterns within pattern-less sequences

Abstract: While Kolmogorov complexity, also known as Algorithmic Information Theory, defines a measure of randomness as being pattern-less in a sequence of a binary string, such rubrics come into question when sub-groupings are used as a measure of such patterns in a similar sequence of a binary string. This paper examines such sub-group patterns and finds questions raised about existing measures for a random binary string. PACS Numbers: 89.70tc, 89.20Ff, 89.70tc, 84.40Ua

Qualities of randomness and non-randomness have their origins with the work of von Mises in the area of probability and statistics [1]. While most experts feel all random probabilities are by nature actually pseudo-random in nature, a sub-field of statistical communication theory, also known as information theory, has developed a standard measure of randomness known as Kolmogorov randomness, also known as Martin-L of randomness, that was developed in the 1960s [2–4]. This sub-field of information theory is known as Algorithmic Information Theory [5]. What makes this measure of randomness, and non-randomness, so distinct is the notion of patterns, and pattern less, sequences of 1's and 0's in a string of binary symbols [6]. In other words, perceptual patterns as seen in a sequence of objects that can be defined as having similar sub-groupings within the body of the sequence that have a frequency, depending on the length of the string, of either regularity, non-randomness, or infrequency, randomness, within the sequence itself [7].

In examining the classical notion of a random and non-random set of 1's and 0's in two examples of a sequence of binary strings, the pattern verses pattern-less qualities can be examined. Example #1 is as follows: [111000111000111] and Example #2 is as follows: [110111001000011]. It is clear than Example #1 is more patterned than Example #2 in that Example #1 has a balanced subgroups of three characters, either all 1's or all 0's, that have a perceptual regularity. Example #2 is a classical model of a sequence of a random binary string in that the sub-groups, if grouped into like, or similar, characters, either all 1's or all 0's like in Example #1, the frequency of the types of characters, either 1's or 0's, is different, seven variations of groups as opposed to the five variations in Example #1, as are the subgroups : [(11), (0), CUD , (00), (1), (0000) , & (11)] from Example #2. While this would support the pattern verses pattern-less model proposed by Kolmogorov complexity, there is a striking result from these two examples, #1 and #2, in that the second, or random, example, Example #2, has a pattern within the sub-groups, that for all perceptual accounts, has distinct qualities that can be used to measure the nature of randomness on a sub-grouped level on examination of a binary string.

The author has done early work on coding each of the sub-groups and reducing them to a compressed state, and then decompressing them with no loss to either the amount of frequency or number of characters to a sequence of a binary string that would be considered random by Kolmogorov complexity [8]. Now, while this simple program of compression and decompression by the author is for a future paper, the real interest of this paper is on the sub-groups as they stand without the notion of compression.

The very idea of the notion of a patterned or pattern-less quality as found in the measure of such aspects to the sub-groupings of 1's and 0's in a sequence of a binary string has the quality of being a bit vague, in that both Example #1 and Example #2 are patterned, in that they have a frequency and similar character sub-groupings that have a known measure and quality that can be quantified in both examples. This is more than a question of semantics as the very nature of the measure of Kolmogorov complexity is the very fact that it has a perceptual 'pattern' to measure the randomness of a sequence of a binary string. In reviewing the literature on the notions of patterns in Kolmogorov complexity/Algorithmic Information Theory the real question arises, which patterns qualify for status as random, especially as a measure in a sequence of a binary string?

References

[1] Knuth, D.E., The Art of Computer Programming: Volume 2 Semi numerical Algorithms (Addi son-Wesley Publishers, Reading), 1997, p. 149.

[2] Knuth, D.E., The Art of Computer Programming: Volume 2 Semi numerical Programming (Addison-Wesley Publishers, Reading), 1997, p. 169–170.

[3] Shannon, C.E., Bell Labs. Tech. Jour. 27, (1948), 379–423 & 623–656.

[4] Li, M. and Vitanyi, P., An Introduction to Kolmogorov Complexity and Its Applications (Springer, New York), 1997, p. 186.

[5] M. Ge, The New Encyclopedia Britannica (Encyclopedia Britannica, Chicago), 2005, p.637.

[6] Martin-L of, P., Infor. And Contr., 9.6 (1966), 602–619.

[7] Uspensky, V.A., 'An introduction to the theory of Kolmogorov complexity' edited by Watanabe, O. Kolmogorov Complexity and Computational Complexity (Springer-Verlag, Berlin), 1992, p. 87.

[8] Tice, B.S., Formal Constraints to Formal Languages (AuthorHouse, Bloomington), in press.

Appendix I: A radix 4 based system for use in theoretical genetics

Abstract: The paper will introduce the quaternary, or radix 4, based system for use as a fundamental standard beyond the traditional binary, or radix 2, based system in use today. A greater level of compression is noted in the radix 4 based system when compared to the radix 2 base as applied to a model of information theory. The application of this compression algorithm to both DNA and RNA sequences for compression will be reviewed in this paper.

Key words: Radix 4, Quaternary, Theoretical Genetics, DNA Compression, RNA Compression

I. Introduction: A quaternary, or radix 4 based system, is defined as four separate characters, or symbols, that have no semantic meaning apart from not representing the other characters. This is the same notion Shannon gave to the binary based system upon its publication in 1948 [1]. This paper will present research that shows the radix 4 based system to have a compression value greater than the traditional radix 2 based system in use today [2]. The compression algorithm will be used to compress DNA and RNA sequences. The work has applications in theoretical genetics and synthetic biology.

2. Randomness: The earliest definition for randomness in a string of 1's and 0's was defined by von Mises, but it was Martin-Lof's paper of 1966 that gave a measure to randomness by the *patternlessness* of a sequence of 1's and 0's in a string that could be used to define a random binary sequence in a string [3 and 4]. A non-random string will be able to compress, were as a random string of characters will not be able to compress. This is the classical measure for Kolmogorov complexity, also known as Algorithmic Information Theory, of the randomness of a sequence found in a binary string.

3. Compression Program: The compression program to be used has been termed the Modified Symbolic Space Multiplier

Program as it simply notes the first character in a line of characters in a binary sequence of a string and subgroups them into common or like groups of similar characters, all 1's grouped with 1's and all 0's grouped with 0's, in that string and is assigned a single character notation that represents the number found in that sub-group, so that it can be reduced, compressed, and decompressed, expanded, back to its original length and form [5]. An underlined 1 or 0 is usually used to note the notation symbol for the placement and character type in previous applications of this program. The underlined initial character to be compressed will be used for this paper.

4. Application of Theory: The application of a quaternary, or radix 4 based system, to existing genetic marking and counting systems has many advantages. The first is the greater amount of compression from this base, as opposed to the standard binary based system in use today, and secondly, as a more utilizable system because of the four character, or symbol, based system that provides for more variety to develop information applications.

5. DNA: DNA, or Deoxyribonucleic acid, is a linear polymer made up of specific repeating segments of phosphodiester bonds and is a carrier of genetic information [6]. There are four bases in DNA: adenine, thymine, guanine and cytosine [7].

The use of a compression algorithm for sequences of DNA.

Definitions A

= Adenine T

= Thymine G

= Guanine C

= Cytosine

Example #

A ATATGCGCTATACGCGTATATATA

The compression algorithm will use a specific focus on TA and GC DNA sequences in Example #A.

Key Code

TA = 4 characters

GC = 2 characters

Compress Example #A

ATATGCATATCGCGTA

The compressed DNA sequence is 16 characters from the original non-compression total of 24.

The use of a four character system, a radix 4 base number system, that is composed of each character not representing the other characters is ideal in DNA sequences composed of adenine, thymine, guanine and cytosine.

Example #D

TAGCTAGCTAGCTAGCTAGCTAGCTAGCTAGCTAGCTAGC

Key Code

TAGC = 10

Compression of Example #D

TAGC

The compressed version of Example #D is 4 characters from the original non-compressed total of 40 characters.

6. RNA: RNA, or Ribonucleic acid, translates the genetic information found in DNA into proteins [8]. There are four bases that attach to each ribos [9].

The use of a compression algorithm for sequences of RNA.

Definitions A

= Adenine C

= Cytosine G

= Guanine U

= Uracil

Example #B

AUAUCGCGAUAUCGCGUAUAUAUAGCGC

The compression algorithm will focus on specific RNA sequences.

Key Code

UA = 4 characters

GC = 2 characters

Compress Example #B

AUAUCGCGAUAUCGCGUAGC

The compressed RNA sequence is 20 characters in length from the original non-compression total character length of 28.

The use of a four character system, a radix 4 base number system, that is composed of each character not representing the other characters is ideal in RNA sequences composed of adenine, cytosine, guanine and uracil.

Example #C

UAGCUAGCUAGCUAGCUAGCUAGC

The use of a universal compression algorithm is as follows:

Key Code

UAGC = 6

Compression of Example #C

UAGC

The compressed version of Example #C is 4 characters from the original non-compressed 24 character total length.

Summary: The compression algorithm used for both DNA and RNA sequences has the power of both a universal compression algorithm, all character length types, and a specific, or target, level of compression.

References

1. Shannon, C.E., A Mathematical Theory of Information", *Bell Labs. Tech. Jour.* 27, 379–423 and 623–656 (1948).

2. Tice, B.S., "The analysis of binary, ternary and quaternary based systems for communications theory", Poster for the SPIE Symposium on Optical Engineering and Application Conference, San Diego, California, August 10–14, 2008.

3. Kotz, S. and Johnson, N.I., Encyclopedia of Statistical Sciences, John Wiley & Sons, New York (1982).

4. Martin-L of, P., "The definition of random sequences", *Information and Control,* 9, pp. 602–619 (1966).

5. Tice, abide.

6. L.C. Lutter, "Deoxyribonucleic acid". In *McGraw-Hill Encyclopedia of science & technology.* McGraw-Hill Publishers, New York, pp. 373–379 (2007).

7. Lutter, abide., p. 374.

8. A. L. Beyer and M. W. Gray, "Ribosomes". In *McGraw-Hill Encyclopedia of science & technology.* McGraw-Hill Publishers, New York, pp. 542–546 (2007).

9. Beyer, abide., p. 542.

Appendix J: A compression program for chemical, biological and nanotechnologies

Abstract: The paper will introduce a compression algorithm that will use based number systems beyond the fundamental standard of the traditional binary, or radix 2, based system in use today. A greater level of compression is noted in these radix based number systems when compared to the radix 2 base as applied to a sequential strings of various information. The application of this compression algorithm to both random and non-random sequences for compression will be reviewed in this paper. The natural sciences and engineering applications will be areas covered in this paper.

Key words: Compression Algorithm, Chemistry, Biology, and Nanotechnology

I. Introduction: A binary, or radix 2 based, system is defined as two separate characters, or symbols, that have no semantic meaning apart from not representing the other character. This is the same notion Shannon gave to the binary based system upon its publication in 1948 [1]. This paper will present research that shows how various radix based number systems have a compression value greater than the traditional radix 2 based system as in use today [2]. The compression algorithm will be used to compress various random and non-random sequences. The work has applications in theoretical and applied natural sciences and engineering.

2. Randomness: The earliest definition for randomness in a string of 1's and 0's was defined by von Mises, but it was Martin-Lof 's paper of 1966 that gave a measure to randomness by the patternlessness of a sequence of 1's and 0's in a string that could be used to define a random binary sequence in a string [3 and 4]. A non-random string will be able to compress, where as a random string of characters will not be able to compress. This is the classical measure for Kolmogorov complexity, also known as Algorithmic Information Theory, of the randomness of a sequence found in a binary string.

3. Compression Program: The compression program to be used has been termed the *Modified Symbolic Space Multiplier Program* as it simply notes the first character in a line of characters in a binary sequence of a string and subgroups them into common or like groups of similar characters, all 1's grouped with 1's and all 0's grouped with 0's, in that string and is assigned a single character notation that represents the number found in that sub-group, so that it can be reduced, compressed, and decompressed, expanded, back to its original length and form [5]. An underlined 1 or 0 is usually used to note the notation symbol for the placement and character type in previous applications of this program. The underlined initial character to be compressed will be used for this paper.

4. Application of Theory: The compression algorithm will be used for the following radix based number systems: Radix 6, Radix 8, Radix 10, Radix 12 and Radix 16. These are traditional radix base numbers from the field of computer science and have strong applications to other fields of science and engineering due to the parsimonious nature of these low digit radix base number systems [6]. The compression algorithm in this paper can be both a 'universal' compression engine in that all members of a sequence, either random or non-random, can be compressed or a 'specific' compression engine that compresses only specific types of sub-groups within a random or non-random string of a sequence.

The compression algorithm will be defined by the following properties:

1. Starting at the far left of the string, the beginning, and moving to the right, towards the end of the string.

2. Each sub-group of common characters, including singular characters, will be grouped into common sub-groups and marked accordingly.

3. The notation for marking each sub-group will be underling the initial character of that common sub-group. The

remaining common characters in that marked sub-group will be removed. This results in a compressed sequential string.

4. De-compression of the compressed string is the reverse process with complete position and character count to the original pre-compressed sequential string.

5. This will be the same processes for both random and non-random sequential strings.

5. Chemistry: Chemistry is the science of the structure, the properties and the composition of matter and its changes [7].

5.1   Polymer

A polymer is macromolecule, large molecule, made up of repeating structural segments usually connected by covalent chemical bonds [8].

5.2   Copolymer

A copolymer, also known as a heteropolymer, is a polymer derived from two or more monomers [9].

Types of copolymers:

1. Alternating Copolymers: Regular alternating A and B units.

2. Periodic Copolymers: A and B units arranged in a repeating sequence.

3. Statistical Copolymers: Random sequences.

4. Block Copolymers: Made up of two or more homopolymer subunits joined by covalent bonds.

5. Stereoblock Copolymer: A structure formed from a monomer.

An example of the use of a compression algorithm on copolymers is as follows:

1. Alternating Copolymers: Alternating copolymers using a radix 2 base number system.

Unit A = 0

Unit B = 1

Example # 1:

01010101010101

Compression of Example # 1

Key Code

0 = 7 characters

1–7 characters

Example # 1 Compressed

01

The compressed state of Example #1 is a 2 character length from the original non-compression state total of 17 characters in length.

Periodic Copolymers: Periodic copolymers using a radix 16 base number system.

Unit A = abcdefghijklmnop

Unit B = 123456789@#$%^&*

Example #2

abcdefghijktamopl23456789@#$%^&*123456789@#$%^&*abcdefghijkhnnopl23456789@#$%^&*

Compression of Example #2

Key Code

abcdefghijklmnop −16 characters

123456789@#$%A&* = 16 characters

Example #2 Compressed

A11a1

The compressed state of Example #2 is 5 characters from the original non-compression state total of a 80 character length.

Statistical Polymers: Random copolymer using a radix 8 base number system.

Unit A =12345678

Unit B = abcdefgh

Example #3

12345678abcdefghabcdefghl234567812345678abcdef
ghl2345678l234567812345678

Key Code

12345678 = 8 characters

abcdefgh = 8 characters

Compression of Example #3

lallalll

The compressed state of Example #3 is 8 from the original non-compression state total of a 64 character length.

Block Copolymers: Block copolymer using a radix 12 base number system.

Unit A = abcdefghijkl

Unit B=123456789@#$

Example #4

123456789@#$123456789@#$123456789@#$abcdefghijklab
cdefghijklabcdefghijkl

Key Code

abcdefghijk =12 characters

123456789@#$ = 12 characters

Compression of Example #4

111aaa

The compressed state of Example #4 is 6 characters from the original non-compression state of 58 character length.

Stereoblock Copolymer: Stereoblock copolymer using a radix 10 base number system.

Unit A = abcdefghij

Unit B = 123456789@

Note: The symbol [I] represents a special structure defining each block.

Example #5

abcdefghij   abcdefghij   abcdefghij   abcdefghij   abcdefghij
abcedfghij

I   I

123456789@123456789@   123456789@123456789@

Key Code

abcedfghij = 10 characters

123456789@ = 10 characters

Compression of Example #5

aaaaaa

1111

The compressed state of Example #5 is 10 characters from the original non-compression total of 100 characters in length.

6. Biology: Biology is the study of nature and as such is a part of the systematic atomistic axiomization of processes found within living things. These natural grammars, or laws, have mathematical corollates that parallel process found in the physical and engineering disciplines. The use of a compression algorithm of a sequential string is a natural development of such a process as can be seen in the compression of both DNA and RNA genetic codes.

6.1   DNA

DNA or Deoxyribonucleic acid is a linear polymer made up of specific repeating segments of phosphodiester bonds and is a carrier of genetic information [10]. There are four bases in DNA: adenine, thymine, guanine and cytosine [11].

The use of a compression algorithm for sequences of DNA.

Definitions:

A = Adenine

T = Thymine

G = Guanine

C = Cytosine

Example #A

ATATGCGCATATCGCGTATATATATA

The compression algorithm will use a specific focus on TA and GC DNA sequences in Example #A.

Key Code

TA = 6 characters

GC = 2 characters

Compress Example #A

ATATGCATATCGCGTA

The compressed DNA sequence is 16 characters from the original non compression total of a 28 character length.

6.2   RNA

RNA, or Ribonucleic acid, translates the genetic information found in DNA into proteins [12]. There are four bases that attached to each ribos [13].

Definitions:

A = Adenine

C = Cytosine

G = Guanine

U = Uracil

Example #B

AUAUCGCGAUAUCGCGUAUAUAUAUAGCGC

The compression algorithm will focus on specific RNA sequences.

Key Code

UA = 6 characters

GC = 2 characters

Compress Example #B

AUAUCGCGAUAUCGCGUAGC

The compressed KNA sequence is 20 characters hi length from the original non-compression total character length of 32.

7. Nanotechnology: The development and discovery of nanometer scale structures, ranging from 1 to 100 nanometers, to transform matter, energy and information on a molecular level of technology [14].

7.1   Synthetic Biology

Within the field of synthetic biology is the development of synthetic genomics that uses aspects of genetic modification on pre-existing life forms to produce a product or desired behavior in the life form created [15].

The following is a DNA sequence of real and 'made up' synthetic sequences.

Definitions:

A =

Adenine T

= Thymine G

= Guanine C

= Cytosine W

= *Watson K

= *Crick

*Note: Made up synthetic DNA.

Example #C

TATAGCGCWKWKATATCGCGKWKWKWKWKWKW

Key Code

AT = 2 characters

CG = 2 characters

KW = 6 characters

Compressed Example #C

TATAGCGCWKWKATCGKW

The compressed synthetic DNA sequence is 18 characters from the original non-compression character total of 32.

Summary: The paper has addressed the use of a compression algorithm for use in various radix based number systems in the fields of chemistry, biology and nanotechnology. The compression algorithm in both the universal and specific format have successfully reduced long and short sequences of strings to very compressed states and function well in both random and non-random sequential strings.

References

1. Shannon, C.E., *Bell Labs. Tech. Jour.* 27, 379–423 and 623–656 (1948).

2. Tice, B.S., "The analysis of binary, ternary and quaternary based systems for communications theory", Poster for the SPIE Symposium on Optical Engineering and Application Conference, San Diego, California, August 10–14, 2008.

3. Kotz, S. and Johnson, N.I., *Encyclopedia of Statistical Sciences,* John Wiley & Sons, New York (1982).

4. Martin-L of, P., "The definition of random sequences", *Information and Control,* 9, pp. 602–619.

5. Tice, abide.

6. Richards, R.K., *Algorithmic Operations in Digital Computers*, D. Van Nostrand Company, Princeton, NJ (1955).

7. Moore, J.A. *McGraw-Hill Encyclopedia of Chemistry*, McGraw-Hill Publishers, New York (1993).

8. Wikipedia. "Polymer". Wikipedia, September 4, 2010, p. 1. Website: http://en.wikipedia.org/wiki/Polymers.

7

9. Wikipedia. "Copolymer". Wikipedia, September 4, 2010, pp. 1–5. Website: http://en.wikipedia.org/wiki/Copolymers.

10. Lutter, L.C., "Deoxyribonucleic acid". In *McGraw-Hill Encyclopedia of science & technology*.

McGraw-Hill Publishers, New York. Pp. 373–379 (2007).

11. Lutter, abide., p. 3–74.

12. Beyer, A.L. and Gray, M.W., "Ribosomes". In *McGraw-Hill Encyclopedia of science & technology*.

McGraw-Hill Publishers, New York. Pp. 542–546 (2007).

13. Beyer, abide., p. 542.

14. Drexler, K.E., "Nanotechnology". In *McGraw-Hill Encyclopedia of science & technology*.

McGraw-Hill Publishers, New York, pp. 604–607 (2007).

15. Wikipedia. "Synthetic genomics". Wikipedia, September 4, 2010, p. 1.

Website: http://en.wikipeida/wiki/Synthetic-genomics.

Appendix K: Statistical physics and the fundamentals of minimum description length and minimum message length

Abstract: This monograph is the first account of the use of a 'summing engine' an algorithm for counting and unifying common, or liked natured, characters in a binary sequential string and combining them into a sub-group of common character types into a compressed collective.

The ability to compress a random binary sequential string is a novel feature of this 'Summing Engine' algorithm and is examined in light of both Minimum Description Length and its progenitor Minimum Message length that use data compression as a parameter for defining 'good' data for evaluating a model for measurement.

A new paradigm for both Minimum Message Length and Minimum Description Length results from this study.

Preface

The application of a counting, or addition, algorithm to compress common, or liked natured, characters in a binary sequential string that reduces, into a compressed state, a random binary sequential string and then utilized in both the Minimum Message Length and the Minimum Description Length models as test systems for the traditional random state and the non-traditional random state as developed by the author in 1998. A new paradigm results changing the fundamental notions in both the Minimum Message Length (1968) and the modern version of the Minimum Description Length (1978) model.

Introduction

The monograph addresses the 'compressibility' of a traditional random binary segmental string verses a 'summing engine' binary sequential string. The model used for both compression systems are the Minimum Message Length, MML, and the Minimum Description Length, MDL, with the result being a paradigm shift of both the Minimum Message Length and to the Minimum Description Length systems at the fundamental level.

Minimum Message Length, MML, was first developed by C. S. Wallace and D. M. Boulton (1968). The bases of Minimal Message Length is very similar to Minimum Description Length except the Minimum Message Length is a fully subject Bayesian model (Wikipedia, 2011: 1 & 4).

Minimum Description Length, MDL, was first developed by J. Rissanen in 1978 (Rissanen, 1978). The basis of the Minimum Description Length is that regularity in a specific set of data can be used to compress the data into a sequence shorter than the original length of the originating sequence.

Minimum message length

Minimum Message Length [MML] was the early progenitor of Minimal Description Length [MDL] and was first published by Wallace and Boulton in 1968. The primary difference between MML and MDL, the Minimum Description Length, is that the Minimum Message Length is a fully subjective Bayesian model in that it is of a 'a prior' distribution (Wikipedia, 2012:5-6).

In the MML model, all the parameters are encoded in the first part of a two-part code so all the parameters are learned (Wikipedia, 2012: 5–6).

Minimum description length

Minimum Description Length [MDL] was developed in 1978 by Jorma Rissanen as a method by which the "best hypothesis for a given set of data is the one that leads to the best compression of the data" (Wikipedia, 2012. 1). Minimum Description Length has had to bypass two fundamentals of Kolmogorov Complexity in that

Kolmogorov Complexity is incomputable and uses 'what' computer language is used (Wikipedia, 2012: 2). Minimum Description Length restricts the codes allowable so that it does become computable to find the shortest code length available and that the code being 'reasonably' efficient (Wikipedia, 2012: 2).

The Minimum Description Length Principle is as follows: The best theory minimizes the bit sum of the length of the description of the theory and the length in bits of the data when encode by the help of the theory (Li and Vitany, 1997: 351). Minimum Description Length tries to balance the regularity and randomness in the data using the best model; the one that uses regularity in the data to compress (Li and Vitanyi, 1997: 351).

The grammar of form

Martin-L of (1966) has developed a form of algorithmic complexity-based on patterns within sequential strings of binary data of both a random and non-random manner (Martin-L of, 1966). A more regular pattern of a binary sequential string would look like this:

[1010101010]

This type of regularity has regularity, a balance, of form of both [0's] and [1's] that marks it as a non-random sequential string (Martin-L of, 1966). A random sequential string would look like the following:

[0100011000]

These patterns represent the patterns of the grammar of statistical randomness and are the tradition measure of 'the qualities and quantities' of the notion of statistical randomness in a sequential string.

A compression engine

A 'compression engine' is, in essence, an algorithm for common types of characters, either all [0's] or [1's], in a binary sequential string to compress into like-natured characters and then be de-compressed when desired (Tice, 2009). Much of the 'ornamental', visual markers, found in my first large scale address of a 'summing engine' is removed from this publication to save time and extraneous notation (Tice, 2009).

The 'summing engine' is a systematic process of the sequential addition of common, or liked natured characters. In this case the binary [1's] and [0's] of minimally weighted semantic values that have the [1's] being the opposite value of the [0's].

A new paradigm

If a 'summing engine' is used on a binary sequential string, the following three results will occur:

The pattern of the binary sequential string will be of a regular pattern or non-random distribution.

The pattern of the binary sequential string will be of a non-regular pattern or of a random distribution.

The pattern of the binary sequential string will be of both a regular and a random distribution.

Along with the three types of binary sequential string types, the fundamental properties of 'a priori' and 'a posteriori' mark the pre-algorithm and the post-algorithm models.

A traditionally compressed binary sequential string would be non-random as follows:

[1010101010]

A traditional binary sequential string would not compress as follows:

[0111100011]

If a 'summing engine' is used as the algorithm, both non-random and random binary sequential strings would both compress as follows: Non-random compression:

Compressed:

[10] five times.

Random compression:

[1000110000] with [lxl] [0x3] [1x2] [0x4] or [1010]

If the 'summing engine' is used along with the traditional notions of Kolmogorov Complexity, a great deal of change is noted for the 'summing engines' ability to compress a random binary sequential string as to the traditional Kolmogorov Complexity random model.

Minimal Description Length avoids assumptions about data generating procedures, where as Minimum Message Length 'represents a Bayesian framework' (Wikipedia, 2012: 4). It has been noted in the literature that some researchers feel that Minimum Description Length is equivalent to Bayesian inference but is renounced by Rissanen as being 'data that does not reflect the 'true' nature of whether the data as collected or that data to 'reality' (Wikipedia, 2012: 4).

Because two forms of randomness exist traditional, unable to compress randomness, and the compressible type of a 'summing engine', able to compress a random sequential string, the very fundamental of both the Minimum Message Length and the more developed model of the Minimum Description Length are divergent at the point of what makes up a 'random binary sequential string and compression'.

Modern Minimum Description Length theory is based on the principle of compression, as is traditionally known Kolmogorov Complexity, and has no development using a random binary sequential string using a 'summing engine' (Wikipedia, 2012: 1 and Li and Vitanyi, 1997: 351).

Conclusion

Using both the Minimal Message Length Model and the Minimal Description Length Model as 'tests' of randomness found in binary sequential strings against both the traditional Kolmogorov Complexity notion of randomness and the author's 'Summing Engine' form of randomness in binary sequential strings result in two very different sets of measures.

The resulting changes to both the Minimal Message Length and the Minimal Description Length models is at the fundamental level of statistical physics and adds a new chapter to the study of algorithmic complexity.

Summary

This monograph is the first account of a major fundamental change to both Minimum Message Length and Minimum Description Length. The application of a 'summing engine' to make a random binary sequential string to compress has foundational effects to the traditional notions about compression as a fundamental level of the parameters for the measure of such compression found in both the Minimum Message Length and the Minimum Description Length models.

Notes

Rissanen notes his, with Barron and Yu, that Wallace and Boulton's seminal paper (1968) to the field of Minimal Description Length as being the "crudest". Wallace and Boulton's paper of 1968 seems more 'rudimentary' than 'crude' and is, in some respects, a different set of measures than Rissanen's work (Barron, Rissanen and Yu, 1978: 1 and Wallace and Boulton, 1968).

References

Barron, A., Rissanen, J., and Yu, B. (1998). "The Minimum Description Length Principle in Coding and modeling". IEEE Transactions on Information Theory, Vol.44, No. 6, pp. 1–17.

Li, M. and Vitany, P. (1997). An Introduction to Kolomogov Complexity and Its Application. Second Edition. New York, Springer.

Li, M. and Vitanyi, P. (2000). "Minimum Description Length Induction, Bayesiansim, and Kolmogorov Complexity". IEEE Transactions on Information Theory, Vol. 46, No. 2.

Rissanen, J. (1978). "Modeling by shortest data description". Automatica, Vol. 14, pp. 465–4711978.

Tice, B.S. (2009). Aspects of Kolomogorov: The Physics of Information. Denmark: River Publishers.

Vitanyi, P. and Li, M. (2008). "Minimum Description Length Induction, Bayyesnism and Kolmogorov Complexity". arXiv;cs/9901014vl [cs.LG], pp. 1–35.

Wallace, C.S. and Boulton, D.M. (1968). "An Information Measure for Classification". Comp. J. Vol. 11, No. 2, pp. 185–195.

Wikipedia (2012). "Minimum Description Length".

Wikipedia Encyclopedia, pp 1–6.

Website: http;//en.wikipedia.org/wiki/Minimum description-length

Index

Index